Virginian Railway Loco

by

Lloyd D. Lewis

1993
TLC Publishing
Rt. 4 - Box 154
Lynchburg, Virginia 24503-9711

Front Cover Illustration: Virginian Railway locomotives of just about any juncture are best illustrated by an H. Reid color slide of an MB 2-8-2 in action. Thus, we have the 430 switching on Norfolk's Sewells Point Yard, in December 1954. In just four months the 1909 locomotive would be cut up for scrap. In 1993, everything in this photo is gone except the earth and sky. This author's friend H. Reid himself went to his reward late in 1992. Through the kind courtesy of his widow Ginny, we present this cover as a tribute to the man we called "H."

Acknowledgments
Special thanks to my TLC partners, Tom & Carolyn Dixon, for doing the layout, production, and encouragement on this volume, to all the photographers whose work is represented herein, and also to Frank H. Dewey of Jacksonville, Florida, a fellow CSXT railroader and railroad historian, upon whose computer this text was written.

Library of Congress Catalog Number 93-60883
ISBN 1-883089-05-0

Typography & Layout by
Tom & Carolyn Dixon

Printed in the United States by
Walsworth Publishing Co.
Marceline, Missouri 64658

Contents

Introduction 1

Section 1 - Steam: All Shapes and Sizes 5

    0-8-0 Switchers Classes SA and SB 6
    4-4-0 American Type, Class EA 8
    4-6-0 Ten Wheelers, Class TA 9
    4-6-2 Pacifics, Class PA 11
    2-8-0 Consolidations, Classes CA, CB, CC, CD 15
    2-8-2 Mikados, Classes MA, MB, MC, MCA, MD 17
    2-8-4 Berkshires, Class BA 28
    Articulated Locomotives, Classes AA, AB, AC, AD, AE, XA, USA, USA, AG 32

Section 2 - Electric Locomotives, Classes EL-3A, EL-1A, EL-2B, EL-C 45

Section 3 - Diesels: Big and Standard, With One Exception 65

No. 507, one of the Virginian's five 2-8-4s, speeds along the well-kept mainline at Vassa (near Suffolk), Virginia, on November 25, 1949, in this classic H. Reid photo.

Introduction

A 30-year-old, 32-ton 0-6-0 numbered 1 and lettered "DEEPWATER" switches the sawmill at Robson, West Virginia, under sunny skies on a morning in May 1903. Crewmen will soon ease these loads of finished lumber five miles down Loup Creek to the other end of the roughly-laid line, the interchange at Deepwater with the Chesapeake & Ohio Railway.

— Fifteen years later, in the summer of 1918, some of the world's most powerful locomotives steam past each other in the 3 a.m. darkness. Scene is adjacent to Elmore, West Virginia steam shop at the yard's east end as Virginian Railway's brand new ALCO 449-ton 2-10-10-2 Nos. 800 and 801 slog eastbound at 7 m.p.h. through Clarks Gap Mountain's first curve on the rear of a "Hill Run." On the westbound track, 403-ton 2-8-8-8-4 No. 700 drifts back into Elmore Yard, its last Hill Run delayed three times on the mountain as the giant Baldwin experimental ran out of steam and stalled—again.

— Only eight years later, on September 18, 1926, Virginian hosts one of its rare public functions as the first electric-powered coal train pulls into Roanoke, Virginia, all the way from Elmore behind one of the new 641-ton three-unit EL-3As, at that time the largest locomotives in the world. Business cars, railroad executives, politicians, and reporters are all there to celebrate and take note of this new milestone in railroad technology and efficiency.

—Then jump almost 50 years to Saturday, July 4, 1976, America's official Bicentennial Day, at Rural Retreat, Virginia. On this Norfolk & Western Railway Bristol Line excursion train from Roanoke, 29-year-old, 196-ton, 2,400 horsepower former Virginian class DE-RS diesel No. 171 leads N&W Bicentennial-painted SD45 No. 1776 into town. Soon the two units run around the train, putting the much more modern unit in the lead for the return trip. The 171, long repainted from its original yellow with black stripes to just plain black, will do a repeat performance with the 1776 from Roanoke to Winston-Salem, North Carolina, the next day. Thereafter, the 171—last of the 25 famous Virginian Trainmasters and practically the last active former Virginian locomotive of any type—will be retired.

* * * * *

From first to last, from beginning to end, the not-quite 600-mile-long Virginian Railway was a real family of people, large profits, a good way to make a living, high-tech transportation, big cars, big trains, and many huge locomotives.

This book will review those steam, electric, and diesel-electric locomotives of all sizes that hauled and pushed Virginian trains from southern West Virginia all the way across Virginia and back. This is the second of my efforts to pull together material gathered for several years on this extremely fascinating but gener-

ally little known Class I coal-hauler. If the demand exists, we will make this a series, we can assure you. There's plenty of material—plus a tremendous desire to search out and share it with you.

* * * * *

Why were Virginian's locomotives so big? The answer lies in two realities: tough geography and coal, quite a heavy raw material. As soon as Virginian was completed in 1909, the company's operating people began setting tonnage and length records for the commodity that always accounted for about 95 percent of what it carried to market: bituminous coal from the Pocahontas and Winding Gulf coalfields of Raleigh, Fayette, Wyoming, and Mercer counties, in southern West Virginia.

Those records were first set with what now seem like small Mikados but which were actually quite large for their time. In just a few short years, trains grew and records grew from 2,000 to 4,000 and later 10,000 tons as the locomotives progressed in weight, length and, most importantly, tractive power.

Beginning in 1909 with the world's first road-service 2-8-2 types (designed by Virginian's own brand new mechanical staff), enginemen recruited to railroading from surrounding farms wrestled with the throttles and Johnson bars (reversing mechanism) and coal scoops of 2-6-6-0s, 2-8-8-2s, 2-10-10-2s, and the one and only 2-8-8-8-4. Only 15 years later, in the mid-1920s, these tough mountaineers literally breathed easier while shoving through tunnels during first runs of the side-rodded and just plain massive 1-D-1+1-D-1+1-D-1 American Locomotive Co.-Westinghouse electric locomotives Nos. 100-111.

Top company officers located in the upper floors of Norfolk (Virginia) Terminal Station kept in touch with division officers in Victoria, Virginia, and Princeton, West Virginia, counting costs and maintaining their physical plant to the highest standards. Such were the results of this long-range planning (plus the effects of the Great Depression) that no additional locomotives were needed from 1926 until after World War II. Then several C&O officers became Virginian's leaders for its last 15 years of independent operation before merger into its Pocahontas Roads competitor, the Norfolk & Western, in 1959.

In those hectic years, the second and third (and last) classes of electrics were purchased, along with eight C&O-design 2-6-6-6s for coal trains east of Roanoke, five C&O-design 2-8-4s for time freight east of Roanoke, seven third-hand N&W 2-8-8-2 Mallets for mine-run service, and 15 0-8-0 switchers at bargain prices from the C&O. Finally, in the early 1950s, Virginian bowed to the reality of a quite old fleet of 33 USRA-design 1919 and 1923 2-8-8-2s, increasing difficulty in buying parts and accessories for them and

Three men who had the greatest impact on the post-WWII Virginian, (left to right) W. R. Coe, Chairman of the Executive Committee, George D. Brooke, Chairman, and Frank Beale, President, about to take an inspection trip in the late 1940s.

this were the 0.6-percent, 9.4-mile grade eastbound from Whitethorne, Virginia, to Allegheny Tunnel at Merrimac, and the 8.5-mile 1.5-percent westbound counterpart on the same mountain from Fagg, Virginia, to Allegheny Tunnel just west of the former resort at Yellow Sulphur Springs. Both these grades have required pushers in years past.

West Virginia, the Mountain State, on the other hand, was an entirely different story. Even the state line crossing in the midst of the East River Bridge near Glen Lyn, Virginia, was on the long grade starting at Oney Gap Tunnel just east of Princeton and ending about 18 miles east at Rich Creek, Virginia.

Generically, railroad grades in West Virginia began at hillside and hilltop coal tipples and eased off (but sure didn't become level) where surveyors said enough dirt could be shoveled to allow a junction with the mainline next to the local creek bed.

As is well known, however, the mainline grade for which the heftiest locomotives were built on New River Division west of Roanoke curved and climbed and bridged and tunneled its way 14 miles from Elmore Yard up the Great Flat Top Mountain to a peak just east of a long hole under the summit at Clarks Gap.

For all its existence, Virginian's most powerful locomotives (with the exception of the AG Blue Ridge class 2-6-6-6s) were built with Clarks Gap's maximum 2.07 percent grade in mind.

Thus, heavy grades and heavy coal accounted for why Virginian's locomotives were relatively so large and powerful. To get more specific, we will present summary essays on the 213 steam locomotives, 56 electric locomotive units and 65 diesels.

The Virginian Railway

The Virginian Railway itself presents a fascinating portrait of a well-run company whose own success by itself may have even precluded it from a long life, corporate mergers being the reality that they are in America.

From its beginning as the independent

the fact that almost every railroad in the country had at least committed to dieselization. Due in large measure to a nationwide survey of railroad motive power practices and diesel builders conducted by George T. Strong, Jr., of the Mechanical Department staff in Princeton, Virginian's top mechanical and operating officers recommended the company buy 1,600-horsepower and 2,400-horsepower diesels from Fairbanks-Morse & Co. of Beloit, Wisconsin.

Big locomotives were needed for the massive coal trains traveling over Virginian's heavy inclines, which were virtually all in West Virginia. Exceptions to

Virginian's AB Class No. 600 was the first 2-8-8-2 compound articulated Mallet type on the railroad, and set the pattern for the purchase of many more of the same type but of ever increasing size, weight, and tractive power. The huge, slow Mallets were ideal for Virginian's heavy coal trains.

Deepwater Railway in 1898 (with its first few miles of train laid as early as 1896), the Virginian went through metamorphosis after metamorphosis in the years until its merger into N&W on December 1, 1959.

Following its operation through a lease by Chesapeake & Ohio from 1898 to 1903, the Deepwater was bought by Henry Huttleston Rogers, a New York financier, and his partners, to develop the coalfields immediately south of the C&O main line in Fayette County, West Virginia.

Intrigue followed intrigue (secret surveys, industry rumors, etc.) and by 1903 the course was set for development of the full Virginian Railway. Rogers and his trusted associates (notably William N. Page of Ansted, West Virginia) were unable to reach agreement with either C&O or N&W officials as to a profitable division of transportation rates for coal proposed to be interchanged between the respective carriers.

Thus, Rogers, realizing the value of his coal properties and being the extraordinary businessman that he was, decided to dig into his own "deep pockets" and pay as much as $40 million of his personal fortune to blast, dig, fill, and bridge a right-of-way across southern West Virginia and across Virginia to his own coal pier at Sewells Point, Norfolk, Virginia, in direct competition with N&W's pier at Lambert's point and C&O's across Hampton Roads at Newport News.

Employing about 10,000 men at one time, the Deepwater built east and the Tidewater built west to meet at the west end of the longest bridge on the line, the 2,155-foot-long structure over the New River at Glen Lyn, Virginia, practically on the West Virginia state line. A last spike ceremony occurred just a few months before official operation of the Deepwater-to-Norfolk main line began on July 1, 1909, one of America's very last major railroads to be built, much to the chagrin of N&W and C&O, whose territory was truly being invaded.

Several West Virginia branches were built to the several dozen mines that the Virginian served during its life. In the final flurry of construction, Virginian entered into its era of big locomotives and big cars, roughly from about 1910 to 1926. In those years, several classes of Mallet articulateds were constructed by this small Class I railroad to haul some of the world's heaviest trains. In addition, more than 2,000 120-ton capacity gondolas were built for mainly on-line Virginian service, gons so big they were nicknamed "battleships" and whose existence for more than four decades was a record for that time.

The big locomotive era ended with electrification of the Mullens-to-Roanoke main line in 1924-26 and purchase of 36 separate units of the world's largest locomotives. The Great Depression resulted in a downturn in earnings and the scrapping of Virginian's oldest, lightest locomotives in the mid-1930s.

World War II resulted in huge increases in traffic on all parts of the line, as had World War I, when the armed forces' demand for more and more coal grew. In the late 1940s, Virginian had its last fling in steam by buying its only venture into Superpower but then dieselized and eliminated the remaining pair of passenger trains in the mid-1950s.

The legal end of The Virginian came at 12:01 a.m., December 1, 1959, as N&W accomplished what it had tried to in 1925 but was rebuffed by the Interstate Commerce Commission: gain complete and final control. But although the corporation is gone, as are the locomotives, cars, and most of the lineside structures (including Princeton Shop as a working facility since June 15, 1991), most of the track is still around.

Major abandonments are the 119.9 miles from Suffolk to Abilene, Virginia, the Sewells Point Yard and coal pier complex, about 10 miles from east of Narrows, Virginia, to Kelleysville, West Virginia for a highway project, and relocation of almost 20 miles of branch line in the Guyandotte River basin west of Elmore for a dam project. Even the double-track over Clarks Gap Grade is cut back to single track east of Herndon. Most importantly, coal shipments are way down from what they were at merger time, with only about two trains of loads and two trains of empties rolling through Princeton each day, for inspection by retired dispatcher and "Princeton Yardmaster" Ken Coleman.

Most everyone who ever worked for the Virginian has gone to his or her heavenly reward. Times change slowly but they sure do change. And this author is so grateful that there are younger folks who still want to read one of the greatest railroad stories ever told.

Virginian's Motive Power Officials

The Virginian Railway Company was, like all entities, many things to many people, e.g., its employees, stockholders, admirers, and historians.

For sure, Virginian stock was a great investment that merged twice into two great investments—into N&W in 1959 and then into Norfolk Southern in 1982. A retired trainmaster friend of this author has a sister who once owned 600 shares of Virginian at the time of the merger into N&W. In 1993 that has yielded several thousand shares of Norfolk Southern after splitting a few times.

Although Virginian always had a great traffic base in West Virginia coal, the business sense to carry out a company's mission must come from strong leadership day after day and year after year. Top Virginian officers came both from within its own ranks and from other railroads, depending on circumstances at the time a need arose.

Presidents, naturally, are crucial to a company's operation but because this book deals with Virginian locomotives, I will briefly mention here the railway's Superintendents of Motive Power (SMP) and the accomplishments of their respective administrations:

— R. P. C. Sanderson was first employed by the Tidewater Railway in Norfolk about the time that Virginian predecessor organized in 1903. He organized the mechanical staff and brought with him from the

The 4-6-2 PA class powered most of the Virginian's few mainline passenger trains. Here No. 211 is at Page, West Virginia, September 20, 1952.

ceremony about 1925 when his daughter officially christened EL-3A No. 100 and inaugurated the new motive power system.

— L. C. Kirkhuff was Virginian SMP during the postwar last-steam fling including the C&O-design Blue Ridge 2-6-6-6s and 2-8-4s and purchase of former Norfolk & Western 2-8-8-2s and ex-C&O 0-8-0s. He also oversaw complete dieselization of the non-electrified portion of the railroad and retired to a Kansas farm at the end of 1956.

— W. W. Osborne was Virginian's last SMP, who received most of the EL-C rectifiers and the last two DE-S Fairbanks-Morse diesel units (Nos. 48 and 49). A graduate of Virginia Polytechnic Institute in Blacksburg, Osborne was a Virginian employee who came up through the shop apprentice program beginning about 1937. Upon the N&W takeover at the end of 1959, Osborne worked in several N&W departments in Roanoke until he retired to his home in Southwest Roanoke several years ago, where he still lives.

As noted in my 1992 book *The Virginian Era*, I urge you to write to me with Virginian stories to tell and photos that we can copy for future works. Address me in care of TLC Publishing, Route 4 - Box 154, Lynchburg, VA 24503-9711.

Lloyd D. Lewis
Atlantic Beach, Florida
September 1, 1993.

Seaboard Air Line Railway a young draftsman named George B. Halstead, who retired from Virginian in 1949 in Princeton as Assistant to the SMP and became the first person other than my father that this author ever talked to deeply about "this fascinating railroad business." Mr. Halstead, who died in 1968, was literally my next-door neighbor as a child.

Mr. Sanderson left the new Virginian Railway to go to work for Baldwin Locomotive Works, probably at the Eddystone Works in Philadelphia. He did not supervise the move of the Mechanical Department from Norfolk to the new shops abuilding at Princeton about 1908 or 1909. A latter-day warranted assumption would be that either Mr. Sanderson or his wife didn't want to move from metropolitan Norfolk to the established but rural town of Princeton, which had a population of about 4,000 people at the time.

— Frank H. Slayton was Virginian SMP from 1910 to 1920 and learned about this job from his brother Clarence, who had been superintendent of construction for the Deepwater Railway with an office at Page, West Virginia, for a few years about 1905. SMP Slayton was administrator of the Mechanical Department during Virginian's "Big Steam Era," of the compound articulated Mallets from 2-6-6-0 AAs, through the XA Triplex and up to AE 2-10-10-2s and US 2-8-8-2s. The thoughts of this man were crucial to what these locomotives looked and acted like. Perhaps looking at a new challenge, Mr. Slayton, who never married, left Princeton to work for Baldwin as a locomotive development official/salesman in Calcutta, India! Quite a change of scenery.

— John William Sasser, SMP of the original Norfolk Southern Railway, came to Princeton as the SMP whose recommendations were vital in Virginian's mid-1920s electrification. This construction and complete remake of Mullens-to-Roanoke train operation was a huge responsibility for many years. But among the perks of Mr. Sasser's family was the Clarks Gap

Section 1 - Steam: All Shapes and Sizes

As a general observation after studying this railroad for several years, the author believes that Virginian researched and designed and then bought and operated the best locomotives that money could buy—bar none. The company, as far as I can determine, hardly if ever made a rash decision about its locomotives, cars, physical plant, or any other part of its operations or functions, including its people.

Virginian's builder Henry Huttleston Rogers certainly started the company off on the right foot by

H. Reid

2-8-4 No. 505 at Algren, Virginia, 1948.

spending all the money necessary to build a truly first-class property through some mighty difficult terrain. Then he and all his successors authorized expenditure of whatever money was necessary to buy the best in equipment, and the human talent to operate it. With all that to start, the company made the profits it was projected to, paid its stockholders and the cycle kept renewing itself: spending money to make money.

This philosophy of buying the best was particularly true in the Age of Steam when locomotives were in no way taken "off the shelf," but were designed by the motive power staffs of both the railroad and the commercial builder (or in the notable cases of Norfolk & Western and Pennsylvania railroads, among others, when the builder was also the railroad—at Roanoke and Altoona respectively).

In its earliest years, Virginian's predecessors

Deepwater Railway in West Virginia and Tidewater Railway in Virginia, each bought a few locomotives from both Baldwin Locomotive Works and American Locomotive Company, probably trying out some from both major builders. Even classes as small as the original SA 0-8-0s had three ALCOs and two Baldwins, the three-locomotive class CA 2-8-0s were ALCO and the two CB 2-8-0s were Baldwin. The first 2-8-2s, the six MA class, were all Baldwin but the first passenger steamers— six EA class 4-4-0s and four TA 4-6-0s— were split, Baldwin and ALCO, respectively. This was through the period 1905-1907.

From 1909 to 1916, Baldwin dominated Virginian's orders but after the failure of the XA-class experimental Triplex (which Virginian apparently did not pay for until it was divided into MD-class 2-8-2 No. 410 and AF-class 2-8-8-0 No. 610), ALCO built all of Virginian's locomotives (except the six PA 4-6-2s but including the ALCO-Westinghouse EL-3A electrics) until after World War II. Then Lima constructed the AG 2-6-6-6s and BA 2-8-4s, GE built the last two classes of electrics and Fairbanks-Morse constructed all diesels save second-hand GE No. 6. But never again Baldwin.

Even in its two ventures into the second-hand locomotive market, the USE-class 2-8-8-2s were ALCO and the SB 0-8-0s were Lima. [The two CD-class 2-8-0s were also built by ALCO but this author counts them not as second-hand purchases, but as being "inherited" from two Raleigh County, West Virginia shortlines that were leased 50-50 by Virginian and Chesapeake & Ohio in 1912.]

Nothing about Virginian was ever very complicated, least of all its locomotive roster, with electrics and diesels being simpler than steam. This Class I railroad (Hard to imagine today a 600-mile-long line could be a Class I!) was a virtual straight line across a state-and-a-half that made lots and lots of money— and made its investors very happy.

John B. Corns

The only surviving Virginian steam locomotive is No. 4, an 0-8-0 switcher, purchased new from Baldwin Locomotive Works in 1910, now on display at the Virginia Museum of Transportation at Roanoke.

0-8-0, Classes SA and SB

The Virginian's five SA-class switchers (with road numbers 1-5, appropriately enough) sufficed for the road's yard work from their purchase in 1909-10 until the SBs were purchased second hand from the C&O in 1950. Three of the class, Nos. 1, 3, and 5, were retired at the end of 1933. No. 2 was sold for scrap in August 1955 and No. 4 given to Princeton, West Virginia, for display in May 1957. Subsequently it was sold to what is now the Virginia Museum of Transportation in Roanoke in 1968, and remains today the only surviving Virginian steam locomotive!

(Right) No. 4 running light near Suffolk, Virginia, on September 8, 1948. Note the stars on cylinder heads and polished number plate, unusual for a lowly switcher.

H. Reid

Road Numbers: 1-3 — Builder: ALCO, Richmond (Construction Nos 45973-74) (May 1909)
Road Numbers: 4-5 — Builder: Baldwin (Construction Nos. 35034-35 (August 1910)
Boiler Pressure: 200psi
Tractive Effort: 45,200 pounds — Cylinders: 22x28 in.
Weights: On Drivers: 182,300 pounds; Tender 111,400
Factor of Adhesion: 4.03
Tender Capacity: 10 tons/5,000 gallons

Lloyd D. Lewis Coll.

Builders photo (above) and Virginian mechanical department diagram (below) show the SA's dimensions and appearance. About the only change in outward appearance throughout their lives seems to have been the replacement of the oil headlights, elimination of the tender coal boards, and addition of handrails.

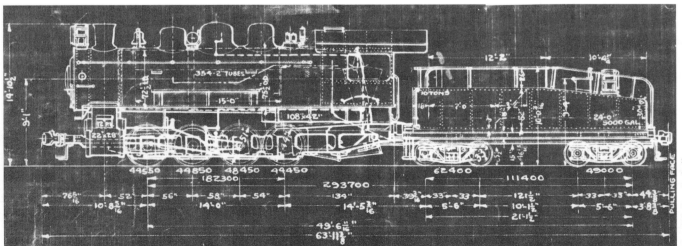

Lloyd D. Lewis Coll.

Class SB - Road Nos. 240-254
Built: Lima, 1942-43 (Construction Nos. 7963-7977) for C&O,
Purchased by VGN September 1950.
Boiler Pressure: 200psi
Tractive Effort: 57,200 pounds — Cylinders: 25x28 in.
Weights: On Drivers: 244,000 pounds; Tender: 158,400 pounds
Factor of Adhesion: 4.26
Tender Capacity: 12 tons/8,000 gallons
[Scrapped 1957-59.]

(Left) SA No. 2 has a single box car in tow at Suffolk January 28, 1954. Although the handrails have been added to the tender, as on No. 4, the built-up coal bunker remains. *(Below)* Head-on view of SB No. 246 new at Lima in 1942. This is when the locomotive was built for C&O, but the face changed not at all when it came to Virginian, retaining even the C&O-style numberboard.

Lloyd D. Lewis Coll.

(Left) SB No. 248 switches under the wires at Roanoke Yard on August 14, 1953. Note the overfire jets on the side of the smokebox, painted aluminum for some unknown reason.

(both) Bob's Photo

Smokily going about its business at Sewell's Point Yard in Norfolk, No. 242 works some loaded hoppers on November 6, 1954. These powerful locomotives could exert 57,200 pounds of tractive effort, more than the MB road Mikes, and ideal for switching heavy cuts of coal cars.

4-4-0 American Type - Class EA

The EA Class 4-4-0s were all built for Virginian predecessor Deepwater Railway by Baldwin in 1906-07. This was a late date for the construction of the 19th Century's premier wheel arrangement, which had been supplanted by the turn of this century almost completely by the heavier 4-4-2 and 4-6-2 types. But for roads with light passenger traffic, such as the nacent Virginian, the 4-4-0 was still an option. One of the six in the class—No. 295—amazingly survived until 1953, still working light passenger jobs (and then was stored at Princeton Roundhouse for several years), while the others all were scrapped in the 1930s. The builder view above shows Deepwater No. 13 (later VGN 103) new in 1906.

William P. Nixon

Road Nos. 100-105 (later 294-299)
Baldwin 1906-07
Tractive Effort: 21,400 pounds
Cylinders: 18x26; Drivers: 63 in.
Boiler Pressure: 200 psi
Weights: On Drivers: 107,500 pounds; Tender: 134,500 pounds.
Factor of Adhesion: 4.62
Tender Capacity: 10 tons/7,000 gallons
Scrapped: No. 292 (6/37); No.294 (8/53); No.296 (6/37); No.297 (11/34); No.298 (11/34); No.299 (6/37)

(Above) No. 294, formerly 100, at Roanoke in the early to mid-1930s with an eastbound passenger train, at the only brick depot structure on the entire railroad. The classic proportions of its boiler and tall stack and domes assign its design to the previous century. *(Below)* Official diagram of the class, revised to 1938.

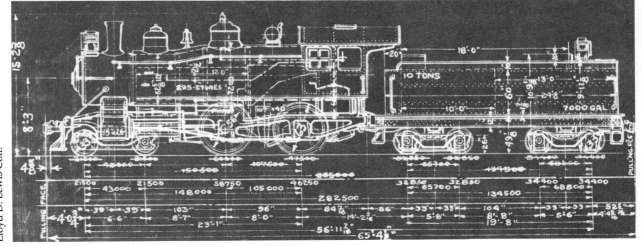

8

4-4-0 No. 296 struts its stuff with a two-car accommodation train at Victoria, Virginia, in the early 1930s. The 67-inch drivers seem small in proportion to this tall locomotive, but two of the class (Nos. 294 and 297) had even smaller ones (63-inch) applied in 1926.

H. W. Pontin, T. W. Dixon Coll.

4-6-0 Ten Wheelers, Class TA

(Above) Typical of many turn-of-the-century Ten Wheelers is the space between boiler and low drivers. Roads that ordered high-drivered locomotives, of course, got a little better proportions. No. 203 was one of four in the class, built by ALCO's Richmond Works in 1907, and lettered Virginian, though built for the Tidewater Railway (the Deepwater and Tidewater lines formed the Virginian). *(Below)* A not entirely clean mechanical diagram nonetheless demonstrates the clinical design of the TA.

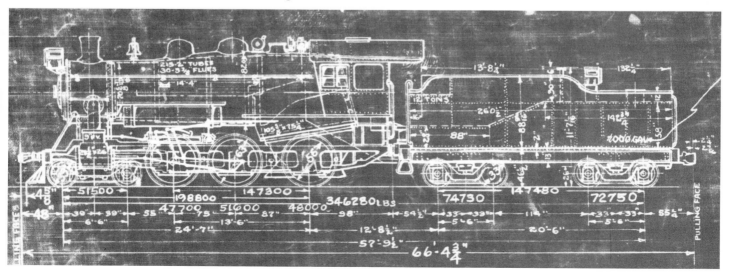

(both) Lloyd D. Lewis Coll.

Road No. 200-203 - ALCO (Richmond), 1907 (Construction Nos. 42966-999)
Tractive Effort:30,900 pounds
Cylinders: 21x26 (Changed to 21.5x26 in 1920 when Southern valve gear replaced Stephenson)
Drivers: 63 inches
Boiler Pressure: 200 psi
Weights: On Drivers: 198,800 pounds
Tender: 147,480 pounds
Factor of Adhesion: 4.54
Tender Capacity: 7,000 gallons (No. 202 tender changed to 8,000 gallons 5/15/45)
Scrapped: No.200,203 in April 1947; No. 201,202 in June 1949.

Like the EA 4-4-0s, Virginian used the TA 4-6-0s everywhere the passenger trains ran at one time or another in their 40-year career. In fact, when brand new, the locomotives were rented for several months during the last half of 1907 to N&W for use west of Bluefield. N&W needed the motive power and Virginian's new track from Deepwater to around Mullens wasn't settled enough for regular traffic yet. Photos of both the TAs and EAs in service are rare because of their early existence in Virginian's history and the lack of photographers especially "over in West Virginia."

H. Reid

TA 4-6-0 No. 202 brings a three-car passenger train into Norfolk Union Depot in 1947, near the end of its long career on the Virginian, its once-clean boiler now adorned with numerous appliances and piping.

A nostalgic scene about 1935 at Mullens, West Virginia, shows the meeting of Train No. 3, powered by Pacific No. 210, while Train No. 14 (left) has TA No. 201 for power. No. 3 continues its daily run from Roanoke west to Charleston, West Virginia, after all the station work is finished. No. 14 has come down the Winding Gulf Branch from Fireco, 31 miles deeper into the hollows than Mullens. This train started the day as No. 13 here at Mullens and is shown completing a daily round trip.

10

T. W. Dixon Coll.

4-6-2 Pacifics - Class PA

Although the Virginian was not a passenger-hauler to any great extent, what short trains it operated were usually powered by the PA class Pacifics built in 1920, at ALCO's Richmond Works. These sturdy locomotives served until the end of passenger service, the trains being discontinued before there was any necessity to convert them to diesel power. The Pacific type was perhaps the most popular design for passenger power when the Virginian bought these six examples. They were average light Pacifics that strove to set no records, just keep the mainline accommodation trains on schedule. Pretty much out of a job after the termination of Virginian passenger service at the end of January 1956, the PAs were surplus and No. 212 was the last to be sold for scrap in January 1960.

Road Numbers: 210-215 (Built by ALCO
  (Richmond) 1920 Construction Nos. 61992-97)
Tractive Effort: 44,300 pounds
Cylinders: 26x28 - Drivers: 69 inches
Boiler Pressure: 200 psi (190 until 1937)
Weights: On Drivers: 189,000 pounds; Tender:
  189,100 pounds
Factor of Adhesion: 4.05
Tender Capacity: 14 tons/10,500 gallons
Scrapped: 1957-60

(both) H. Reid

(Above) No. 210 speeds along under the wires of electric territory near Low Gap Bridge west of Princeton, West Virginia, on August 10, 1948. The plain face of early Twentieth Century motive power is readily visible as 210 charges westward on Train 3, the only concession to more modern practice is the small electric headlight.

(Right) PA No. 214 wheels the eastbound mainline passenger train No. 4 at Tidewater Tower, Norfolk, Virginia, on November 5, 1947, popping noisily from behind a berm, working only slightly with its three-car charge. The VGN had few interline passengers, most of its patrons being folks who lived in towns along its West Virginia-Virginia line that needed to move about only on relatively short trips. A rather large but undocumented percentage of passengers were annual pass-holding employees on assignment and their families on day trips, both groups known universally as "deadheads."

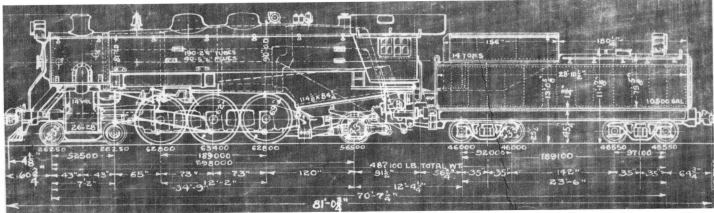

(both) Lloyd D. Lewis Coll.

The builder's broadside photo and mechanical diagram, show the PA's compact and pleasing, conventional design typical of its 1920 production. Over the many years service, the engines' appearance charged hardly at all. Virginian patronized ALCO's Richmond Works in its early period for several orders, giving the business to a Southern builder when possible, as did other roads south of the Mason-Dixon line. ALCO closed its Richmond Works in 1926 after building some of the finest looking Pacifics of all time, the Southern Railway PS-4 and the C&O F-19 classes.

Here No. 215 is at Roanoke in September 1949, with westbound three-car Train No. 3 completing its Norfolk-Roanoke trek.

George E. Votava, T. W. Dixon Coll.

PA No. 214 takes No. 3 out of Norfolk, at Monsanto, Virginia, on June 14, 1949, making what appears to be pretty good speed. One could imagine a considerable train behind the visible cars, but in fact it was only three cars, the first usually including a 30-foot RPO "apartment." This low-angle view shows the PA's compact design to good advantage.

Train No. 3 leaves Roanoke westbound on October 8, 1937, with PA No. 211 sooting up the overhead wires.

13

J. R. Quinn

In its very last days, No. 212 is shined up, its tires and running board in white and ready for viewing by the attendees at the national convention of the National Railway Historical Society, held in Roanoke the first weekend of September 1957. Already out of service for over a year, No. 212 was the last of the six PAs to be sold for scrap, in January 1960. What a way to go out, with spit and polish for railfan cameras!

A sad day at Victoria, Virginia, as the last runs of Nos. 3 and 4 meet, and many people make a last nostalgic trip on January 29, 1956. PA 213 has No. 4 on this last occasion. Note the man standing just to the right of the locomotive, with his child in his arms. Will that youngster remember this day of steam and steel, or understand its importance?

2-8-0 Consolidation Type - Classes CA, CB, CC, CD

Builder's photo of Virginian Consolidation No. 305, new at ALCO's Richmond Works in April 1909.

Class	Road Nos.	Built	Cyl/Drivers	T. E.	Wt. on Dr.	B.P.	Tender Wt/Cap.
CA	300-302 [a]	ALCO/1904-05	20x24/50-in.	29,400	119,425 pounds	180psi	105,540/7 tons/6,000 gal. [b]
CB	303-304 [c]	Baldwin/1905	20x24/50-in.	29,400	120,500 pounds	180psi	119,000/7 tons/6,000 gal.
CC	305	ALCO/1909	20x24/50-in.	29,400	122,300 pounds	180psi	105,500/7 tons/6,000 gal.
CD	306-307 [d]	ALCO/1907	20x24/50-in.	29,400	122,650 pounds	180psi	117,000/7 tons/6,000 gal.

Notes:
[a] Originally Deepwater Ry No. 2, Tidewater Ry Nos. 3,4.
[b] No.300 had tender capacity of 7 tons but 4,500 gallons, and weight of 90,980 lbs.
[c] Originally Tidewater Nos. 5,6.
[d] No. 306 originally White Oak Ry No. 98 (to VGN in 1912); No. 307 originally Piney River & Paint Creek Ry No. 199 (to VGN in 1912).

The Virginian came too late on the scene of American railroading to be heavily involved with 2-8-0 types, which were the prime general freight locomotives from the mid-1880s until about 1910. The Virginian's primary commodity, coal, was quite heavy and demanded the most powerful locomotives, so when the Mikado (2-8-2) emerged on the scene as a road freight locomotive in 1909, the Virginian naturally opted for it, and, of course, the new Mallets. This left the Consolidations that had been purchased by the Deepwater and Tidewater in their formative years as the few representatives on the roster, along with two inherited from West Virginia coal short lines absorbed in 1912. All the 2-8-0s were scrapped in late 1933 as part of a Depression-era purge of unneeded motive power.

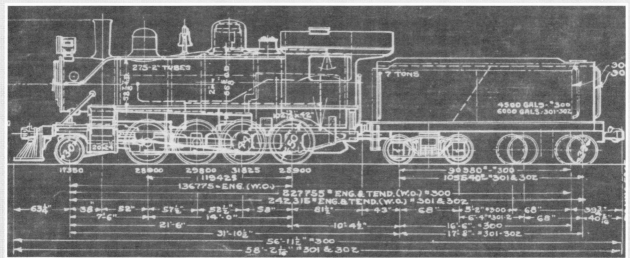

This diagram shows the CA-Class 2-8-0s, built as Deepwater No. 2 and Tidewater Nos. 3 and 4, in 1904-05. They received Virginian Nos. 300-302 in 1907 and were scrapped in 1933.

The Virginian inherited White Oak Railway No. 98 in 1912 when that short line in Raleigh County, West Virginia, coal country, was leased by Virginian and competitor C&O. No. 98 became Virginian 306 in Class CD, working until retirement and scrap in 1933. Here at Lochgelly, West Virginia, White Oak No. 98 works away, miners' clapboard houses on the hill behind. The sign under the cab window reads: "Lahey & Williams Locomotive Fireman Life-Saving Device/ Patent Applied For"—a now-forgotten invention of some type, perhaps having to do with the chains hanging from the cab roof overhang.

Photos of Virginian 2-8-0s are very rare. Here is CA Class No. 302 at Princeton, West Virginia, awaiting scrapping in the early 1930s. Originally Tdewater No. 4, it became Virginian No. 302 in 1907 and held that designation until the end.

(all) Lloyd D. Lewis Coll.

Crewmen, some sporting uniform caps, pose with Piney River & Paint Creek Railroad 2-8-0 No. 199, which became Virginian Class CD No. 307 in 1912. Like the White Oak, the PR&PC was another Raleigh County, West Virginia coal country short line that was jointly leased by the Virginian and C&O in 1912, when No. 199 was purchased by the former. In 1917 the line was absorbed by the C&O entirely, and the White Oak was merged with the Virginian. PR&PC ran between Cranberry, Skelton, Sprauge, Beckley, and Beckley Junction, and tapped developing mines in a region that was soon criss-crossed by a maze of C&O and Virginian branches.

2-8-2 Mikado Type - Classes MA, MB, MC, MCA, MD

The backbone of the Virginian's non-coal Norfolk Division freight fleet and pride of many an engineer was the famous MB class 2-8-2. The 2-8-2 was the most numerous type on the Virginian, and the MB the largest single locomotive class.

Although the 2-8-2 wheel arrangement had been in existence since the early 1880s, few American railroads had purchased the locomotive, and none for genuine road service, before the Virginian. The name that came to be applied to the type originated with Samuel Vauclain of the Baldwin Locomotive Works when he christened 20 2-8-2s being built for the Imperial Japanese Railways as "Mikado Type" in honor of the traditional title of the Japanese emperor.

When the Deepwater and Tidewater Railways went looking for a standard line-haul freight locomotive in 1905 the 2-8-2 was tried, with two locomotives, Deepwater Nos. 30 and 31 (later Virginian 400-401). These locomotives exerted 45,200 pounds of tractive effort and had 174,400 pounds of weight on their driving wheels, making them over 30 per cent heavier and more powerful than the Consolidations then being used. In 1907 four more 2-8-2s of basically the same dimensions were added, and when the Virginian was completed and opened for its full length in 1909, it began receiving what is generally considered the first road-service 2-8-2 of modern configuration with the Class MB, beginning in May of that year and extending through August 1910. A total of 41 locomotives were built, all by Baldwin. An additional 18 in Class MC arrived in 1912.

By 1911, American railroads were thronging to the 2-8-2 type to replace their aging and now outmoded Consolidations. Eventually, more Mikado types were built for use in the United States than any other wheel arrangement, and the Virginian holds an important place in recognizing at the outset the value and utility of this old but now revitalized design.

The Mikado, with its additional trailing truck, allowed the support of a wider and heavier firebox and began the revolution in locomotive design that would eventually led to fireboxes slung entirely behind rather than over or between the drivers. This also resulted in the development of huge locomotives capable of producing steam in large enough quantities to outperform any locomotive that did not have such an innovation.

Althoug the MB was the cost common, the MC was the heaviest and most powerful of the Virginian Mikes, with weight on drivers of 252,100 pounds and tractive effort of 60,800. The MD class was the most unusual, having been created when the Virginian's experimental and unsuccessful 2-8-8-8-4 was broken up into two locomotives, the MD class No. 410 and the AF class No. 610 2-8-8-0.

The five class locomotives, Nos. 480-484, were the finest refinement of the 2-8-2 wheel arrangement by Virginian—and they were some of the relatively few rebuild jobs that Princeton Shop did. The MCA upgrade of five MC—Nos. 472, 466, 470, 475, and 463, in that order—was also done for a specific reason—powering time freights Nos. 71, 72, 73, and 74 on the Norfolk Division: four locomotives and a spare. They were replaced just a few years later by the sleak new BA 2-8-4s.

With parts supplied by Baldwin, including new disc drivers and lightweight side rods, Princeton Shopsmen better counterbalanced the locomotive's motion and added the other touches in the MCA's rehabilitation. The work was done mainly in 1937 and 1941. (Incidentially, the one AF 2-8-8-0 which became a 2-8-8-2 with a trailing truck addition in 1942, the six USE 2-8-8-2 rebuilds in 1935, and the five USD 2-8-8-2 convresions in 1947, 1948, and 1951 were the only similar rehabilitations Princeton Shops undertook in the steam era.)

Overall, Virginian's 66 Mikados were the smaller workhorses of the road's steam power, while the 42 USRA-design 2-8-8-2s were the true larger workhorses. The 42 MBs, particularly, held sway as the predominent presence of any locomotive class throughout the Virginian's history. They were among the very first to go into service on the fledgling Class I, and the 432 was the last Virginian steam locomotive run in Norfolk and probably on the entire Norfolk Division, in the fall of 1956.

Class	Road	Built	Cyl/Drivers	T. E.	Wt. on Dr.	B. P.	Tender Wt/Cap.
MA	400-05	BLW/1905-07	22x28/51-in.	45,200 lbs.	174,400 lbs.	200psi	[c]
MB	420-61	BLW/1909-10	24x36/56-in.	56,000 lbs.	218,500 lbs.	200psi	171,000/16 tons/9,000 gal. [d]
MC	462-79	BLW/1912	26x32/56-in.	65,700 lbs.	252,100 lbs.	200psi	203,300/15 tons/12,000 gal.
MCA	480-84	[a]	26x32/57-in.	64,500 lbs.	250,800 lbs.	200psi	202,200/16 tons/11,500 gal.
MD	410	BLW/1921[b]	26x32/56-in.	65,700 lbs.	233,810 lbs.	200psi	205,140/15 tons/12,000 gal.

Notes: [a] Class MCA was created by Princeton Shops rebuilding five class MC locomotives, Nos. 472, 466, 470, 475, and 463 became 481-484, in that order. —— [b] Class MD consisted of only one locomotive, which was created by Baldwin when the Virginian's 2-8-8-8-4 Triplex was broken apart to become a Mike and a 2-8-8-0 in December 1921. —— [c] Nos. 400-01 had 10 tons/6,000 gallons capacity, while Nos. 402-05 had capacities of 12 tons/7,000 gallons. —— [d] - Hand fired locomotives had tenders weighing 176,000 pounds, and containing 17 tons of coal and 9,500 gallons of water. —— [e] MAs were scrapped in the mid-1930s; eight MBs were scrapped or sold to other roads in the late 1930s, the balance, along with the MCs, MCAs, and MD, were scrapped in the mid-late-1950s.

Rare, indeed, are photos of the Virginian's first Mikados, the MA class. Above is Deepwater No. 30, new at Page, West Virginia, very soon after its arrival in 1905, in a photo taken by the road's superintendent of construction. It later became Virginian No. 400 and was scrapped in 1933. This is the locomotive that originated the 2-8-2 type on the Virginian, and helped establish the Miakdo as a road-service locomotive in the United States.

Waiting for the end at Princeton Roundhouse in 1933 is MA No. 403 which began life in 1907 at Baldwin's Eddystone Works, and served a quarter century for the Virginian before becoming obsolete as a result of declining traffic during the Great Depression.

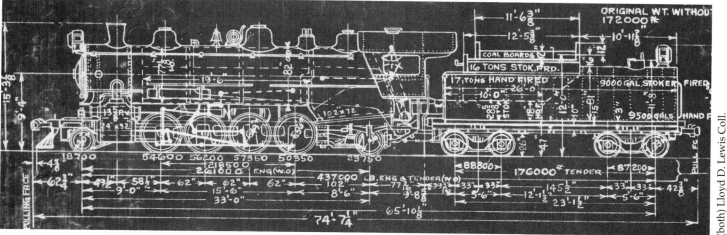

(both) Lloyd D. Lewis Coll.

The Baldwin builder's photo of MB No. 421 and mechanical diagram reveal a large-boilered, low-drivered locomotive, with its firebox almost entirely behind the drivers, supported by its trailing truck. This became more or less the typical Mikado design for the next few decades as the type gained widespread acceptance.

H. Reid

In later life No. 421 at Norfolk on Febrruary 15, 1948, with the Southern Branch Local. Several cosmetic changes have occurred between the photo at the top of this page and this one.

MB 2-8-2 No. 421 is again seen at Sewells Point Yard in Norfolk, Virginia, on May 11, 1947, getting coal from the terminal's chutes. The old wooden water tank at left probably dated the same period as the locomotive itself, both still serving proudly. No. 421's footboards show it to be confined to yard and local work. Basically, the MBs were the veteran Virginian workhorse for as long as the road had steam and almost for as long as the Virginian existed. In fact, No. 459 was built in August 1910 (3-1/2 years after Virginian was formed) and was scrapped in February 1959, nine months before the Virginian also was ended.

(both) H. Reid

Another, wider view of Sewells Point terminal shows several MBs ready for servicing.

H. Reid

MB No. 446 speeds along with a merchandise freight at Norfolk on November 5, 1951.

In the twilight of steam, MB No. 459 backs light at Norfolk on November 6, 1954. Nos. 459 and 432 (built in August and January 1910, respectively), were the last two of the 42 MBs to be scrapped when they were cut up in Februray 1959, only a few months before the Virginian itself merged into N&W. Yard switching and local freight service were common tasks for MBs. Back in 1910 it was the MBs that established the 2-8-2 wheel arrangement as the standard American freight hauler, displacing the 2-8-0. In the following quarter century over 10,000 locomotives of the Mikado type were built, but 1954 there were few steam locomotives of any type left in operation in the United States.

In an unusual backward move MB No. 448 pulls a solid Pullman troop train at the Monsanto Chemical Co. plant in Norfolk on November 18, 1951. Troops movements were, of course, a big thing in the Hampton Roads area where there was a large concentration of Army, Navy, and Air Force bases.

Under the wires at Princeton, West Virginia, No. 443 runs light through the yard in July 1949. The MB is probably the shop and yard switcher on this day.

MB No. 421, a regular at Norfolk, trundles a merchandise train across the South Branch of the Elizabeth River in 1949. This cut of cars could be the Southern Branch Local. Within a year Virginian's ex-C&O 0-8-0 class C-16 (Virginian class SB) were assigned to this job but were unable to keep up with the work. Thus MBs reassumed this role until diesels arrived about 1954.

(both) H. Reid

Photographer H. Reid, premier chronicler of the Virginian in word and photo, captured the one and only MD class, No. 410, riding the turntable at Sewells Point in Norfolk in January 1949. The MD was created when the Virginian's unsuccessful 2-8-8-8-4 Triplex was broken into two locomotives in 1921.

MB No. 432 is seen here switching Sewells Point Yard in 1954. The Mikes were a versatile locomotive, throughout their lives being used for fast freight, coal trains, switching, and even special passenger moves, all illustrated on the preceding pages. Note the coal hoppers in this train are both Virginian and New York Central. The latter have probably come across the big Kanawha River bridge at the far west end of the Virginian at Deepwater, where it connected with the NYC's Kanawha & Michigan subsidiary.

MB No. 446 has a freight near Norfolk Union Station in June 1950. The decorative, though utilitarian, handrails that adorn the otherwise plain face of the locomotive were a distinctive feature. The builder plate appears to have been fairly shined. With its ordinary smokebox, small, high headlight, and footboards, No. 446 would win no beauty contests, but it could move the freight, and that's what counted on any railroad.

(both) H. Reid

Rolling along between Norfolk and Roanoke in relatively flat territory, MC Mikado No. 462 carries white flags this August 15, 1953, as an extra freight hauling a variety of goods. The MC class was rated at 60,800 pounds of tractive force and came to the Virginian in 1912, rounding out its roster of 2-8-2s. No. 462 was eventually scrapped in June 1955.

Bob's Photo

(Below) On October 15, 1937, MC No. 466 is shining and ready for work at Roanoke ready track, its smokebox and firebox stylishly painted in graphite. This is a portrait of a workhorse ready for a hard day's train hauling.

R. P. Morris, Lloyd D. Lewis Coll.

25

MC No. 477 moves a freight at Harmco, West Virginia, 1-1/2 miles west of Mullens. This run is probably Monday-Wednesday-Friday local freight No. 65 beginning its leisurely one way six-hour westbound trip to Page. It will return to-morrow on No. 66.

Aubrey Wiley Coll.

MC No. 464 switches at Princeton on July 24, 1953. Behind it is a gondola loaded with compressed bed-steads, and other scrap from S. S. Belcher & Co.'s scrap yard.

H. Reid

No. 479's crew is all evident in this photo at Elmore Yard coal dock as the MC class 2-8-2 moves out to begin yard duties in June 1940. The brakeman walks ahead to throw a nearby switch to al-low access to the east end of the yard crammed with eastbound coal loads.

Stephen P. Davidson

MCA No. 484 at Victoria, Virginia, on May 31, 1953. The MCA class was created by rebuilding selected class MCs at Princeton Shops during the period 1937-1941. One outward manifestation of the rebulding program was the addition of Baldwin disc drivers. No. 484 resulted from the rebuilding of No. 463 in May 1941. The MCAs carried 200 pounds boiler pressure, which increased their tractive effort ratings over the MCs by 3,300 pounds (64,500).

Another MCA, No. 483 is at Roanoke in April 1955, near the end of the line. It was a March 1941 Princeton rebuild, starting as MC No. 475. A total of five MCAs were created for use on Norfolk Division time freights Nos. 71, 72, 73, and 74, They were supplanted on those trains by the BA 2-8-4s in 1946. The jumble of piping around the boiler and firebox was characteristic of older steam locomotives on most railroads, since they constantly had new appliances and devices added, necessitating additional plumbing that was not planned for in the original design.

2-8-4 Berkshire Type - Class BA

Lloyd D. Lewis C

(Above) Lima builder broadside view. The BA was an almost exact copy of the Chesapeake & Ohio's class K-4. *(Right)* Smokebox view of No. 505 at Roanoke on August 6, 1958. After only ten years the mighty BAs were retired in favor of the diesels. *(Bottom)* Official diagram of the BAs shows, as does the photo above, the compact, powerful design of these locomotives.

Bob's Photo

Road Numbers: 505-509
Builder: Lima - June-July 1946
Bldr. Order No.: 1194
Construction Nos.: 9107-9111
Cylinders: 26x34 inches
Drivers: 69 inches
Weights:
      On Drivers: 295,600 pounds
      Total Eng.: 460,400
      Tender: 410,200
Tractive Effort: 69,350 pounds
Tender Capacity: 21 Tons/25,000 gallons
Boiler Pressure: 245 psi
Firebox: 135-1/16x96-1/4 inches
Total Heating Surface: 4773 sq. ft.
Cylinder Horsepower: 2979
Factor of Adhesion: 4.26
Stoker: Standard HT
Superheater: Elesco Type E

Disposition: All sold for scrap January 1960.

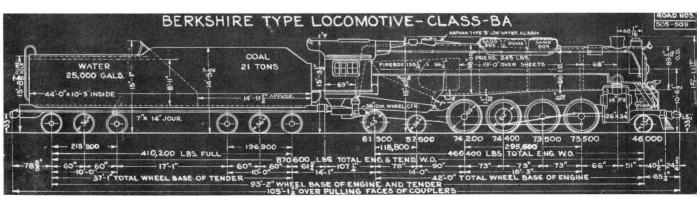

Lloyd D. Lewis C

Right side opaqued builder's photo of No. 505 posed with rods down new at the Lima plant in June 1946. A classic Superpower locomotive from Lima at the height of technological development of the steam locomotive in America.

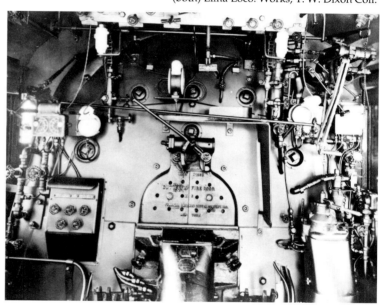

Another Lima builder's view shows the backhead of No. 505. The screw from the type "HT" Standard Stoker can be seen going into the firebox just below the butterfly doors. The tenders on the BA held 21 tons of coal and 25,000 gallons of water, second only to the the AG 2-6-6-6's tenders in capacity on the Virginian.

No. 507 with a time freight train gathers speed at South Norfolk on November 16, 1952. The BAs were a superb general purpose locomotive that could handle heavy trains at speed, run at passenger speeds, and traverse relatively light lines, but on the Virginian they were mainly used for time freights Nos. 71-74. These five locomotives could have been used on coal trains occasionally but the author would like to know if any record exists of their use on passenger or inspection trains. The engineers thought the BAs were great machines.

H. Reid

A Lima Locomotive Works official photo shows No. 505 shining new just outside the erecting shop. This was an "off the shelf" design for Lima which had built 10 almost identical locomotives for C&O in 1945, and would build another 10 in 1947. The C&O eventually had 90 of this type, whereas the Virginian had only the five from 1946.

Lima Loco. Works, Lloyd D. Lewis Coll.

When George Brooke left his position as C&O President to become Virginian Chairman, bringing with him C&O Vice President Frank Beale to be Virginian President, he had already seen the C&O's first 2-8-4s and 2-6-6-6s, thus it was easy for both these top officials to appreciate their designs. Performance on the C&O during WWII was certainly enough to convince even the most casual observer that this particular design was powerful, fast, and highly versatile, capable of assignments on manifest freight, heavy coal, or lightly traveled branches (the C&O even used them in passenger service). It was natural, then, for the Virginian to copy the designs, saving the engineering effort of creating entirely new ones. The BAs, then, were the final outgrowth of a series of designs that began with the Erie 2-8-4s S-class, evolved into the Nickel Plate S-class, the Pere Marquette N-class, and the C&O K-4 class—an impressive pedigree. They were considered "Superpower" locomotives, and as such were in company only with the huge AG 2-6-6-6s on the Virginian. Superpower was a concept pioneered by Lima that embodied a locomotive of large boiler capacity, steamed freely, and included application of the most modern appliances. On the Virginian the BAs ran exclusively east of Roanoke, often hauling freights between there and Victoria in under three hours. On coal trains, they took 65-car, 3,500-ton trains east and returned with 125-car 4,500-ton trains west. The BAs remained in service until about 1955, being replaced by 1,600-horsepower Fairbanks-Morse diesels.

S. K. Bolton, H. H. Harwood, Jr. Coll.

White exhaust flying, BA No. 506 charges along near Suffolk, Virginia, with a freight train in January 1953. The BAs were capable of generating high horsepower at medium speed, and were a very capable general purpose locomotive.

At Algren, Virginia, 15 miles west of Norfolk, BA No. 508 charges along with heavy exhaust, on a merchandise train in 1948. The railroad on the right is Seaboard Air Line's Richmond-Jacksonville main line.

BA No. 507 throws up white exhaust on a cold December 23, 1951, with a coal train at Cottage Toll Road, Norfolk County, Virginia. BAs were usually time freight locomotives, but they could shoulder coal trains and empties with ease. The C&O's design that they copied was for a very powerful all-purpose locomotive.

In September 1957, BA No. 507 is on display beside the Roanoke Motor Shed for the National Railway Historical Society Convention. EL-2B electric No. 126 is at left and boxy EL-3A No.110 is in the background to the right.

31

Articulated Locomotives -
Classes AA, AB, AC, AD, AE, AF, XA, USA, USB, AG

Early in its life the Virginian purchased articulated locomotives, buying strictly compound types except for the final class, the AG. The compound articulated locomotive began to gain favor in the United States about the time the Virginian was completed. It consisted of one large boiler that supplied high pressure steam to a set of cylinders at the rear that powered a set of drivers, then exhausted the partially expanded steam into larger low-pressure cyliders at the front, where it was used again to power a second set of drivers. By using the steam twice, a great deal more energy could be extracted from the same amount of fuel burned, and by using two engines under one boiler, a swivel or articulation joint could allow a much larger boiler to be carried—the rigid wheel base would be only that of each set of drivers and cylinders (engine) under it. Thus, a 2-8-8-2 could round the same curves as a 2-8-0 because its rigid wheelbase was the same (extra clearance, of course, was needed for the over-hang of the boiler as the front engine swiveled).

Virginian bought its first Mallets (as these compound articulateds were called, after their French inventor—Anatole Mallet), in 1909 and assigned them class AA. This 2-6-6-0 wheel arrangement was repeated in class AC in 1910, the same year the single AB class 2-8-8-2 arrived. The 2-8-8-2 type was repeated in the seven ADs of 1912-13, and the USA, USB, USE classes of 1919-23. Nearly unique with the Virginian were the AE 2-10-10-2s of 1918, and certainly unique was the famous and ill-fated XA class 2-8-8-8-4 Triplex. The superpower 2-6-6-6 simple AGs were exact copies of the C&O's H-8 class—only the C&O and the Virginian ever rostered this wheel arrangement (the C&O with 60 and the Virginian with 8). The Mallets were superb for the slow coal train and heavy mine-run switching duties that were the Virginian's life-blood, so they were perfectly suited to their work on this railway.

Class	Road No.	Built	Cyl/Drivers	T.E.	Wt. on Drivers	B. P.	Tender Wt/Capacity
AA	500-503	ALCO/1909 [1]	22x35x30/54-in.	84,960Simple 70,800Comp. [2]	279,900 lb.	200psi	172,000/15 tons/ 4,500 gallons
AB	600	BLW/1910	26x40x32/56-in.	116,900Simple 94,400 Comp.	441,080 lb.	210psi	172,000/15 tons/ 4,500 gallons
AC	510-517	BLW/1910	25x36x32/56-in.	107,500Simple 90,000Comp.	356,780 lb.	210psi	172,000/15 tons/ 9,500 gallons
AD	601-606	ALCO/1912-13	28x44x32/56-in.	138,000Simple 115,000Comp.	479,200 lb.	200psi	204,000/16 tons 12,000 gallons
XA[3]	700	BLW/1916	34(6)x32/56-in.	199,560Simple 166,600Comp.	487,390 lb.	200psi	304,730 lbs./12 tons/ 13,000 gallons
AE	800-809	ALCO/1918	30x48x32/56-in.	176,600Simple 147,200Comp.	617,000 lb.	215psi	214,300 lbs./12 tons/ 13,000 gallons
USA[4]	701-720	ALCO/1919	25x39x32/57-in.	121,600Simple 101,300Comp.	478,000 lb.	240psi	209,100 lbs./16 tons 12,000 gallons
AF[5]	610	BLW/1921	28x44x32/56-in.	138,000Simple 115,000Comp.	487,390 lb.	200psi	222,000/16 tons 13,000 gallons
USB[7]	721-735	ALCO/1923	25x39x32/57-in.	121,600Simple 101,300Comp.	478,000 lb.	240psi	209,100 lbs./16 tons 12,000 gallons
USE[8]	736-742	ALCO/1919	25x39x32/57-in.	136,985Simple 114,154Comp.	491,000 lb.	270psi	290,450/23 tons 16,000 gallons
AG	900-907	Lima/1945	22.5(4)x33/67-in.	110,200	495,000 lb.	260psi	442,000 lbs./25 tons 26,500 gallons

Wheel Arrangements:
Classes AA, AC were 2-6-6-0 (Compound)
Classes AB,AD, USA, USB, USE were 2-8-8-2 (Compound)
Class XA was 2-8-8-8-2 (Simple)
Class AE was 2-10-10-2 (Compound)
Class AG was 2-6-6-6 (Simple)

Notes:
[1] AA, AD, USA, USB at Richmond Works, AE and USE at Schenectady Works.
[2] Value given for compound and simple operation. All Mallets had capability of supplying high-pressure steam directly from boiler to all cylinders, but only for short periods, usually for starting; ran most of the time compound, using steam twice.
[3] XA 2-8-8-8-4 rebuilt by Baldwin in 1922; rear engine and most of boiler becoming AF No.610, and front engine becoming Mikado (MD) No. 410.
[4] Nos. 701-03, and 705 rebuilt by Princeton Shops to class USD in 1947-48 by applying one-piece low-pressure engine beds with cylinders cast integrally.
[5] Created from XA No. 700 in Feb. 1922
[7] Nos. 721, 723, 726,728, 729, 733, rebuilt to class USC in 1935-36 by changing boiler pressure to 250psi, yielding a tractive effort compound of 105,800 pounds. No. 735 included in the USD rebuilt program (see note 4 above) and rebuilt in 1951.
[8] USE class originally N&W Y-4 class. N&W sold locomotives to AT&SF in 1943, which then re-sold them to VGN in 1947.

Disposition (Dates cut up or sold for scrap):
AA, AC in 1933; AD in 1934; AB in 1937; AF 1953; USA in 1953-55; USB in 1954-55; USE in1955; AE in 1948-52; AG in 1960.

Builder photo of Class AA No. 500, the first Virginian Mallet, shows a locomotive of fairly conventional design, looking very much like an overgrown, elongated Consolidation, with a prominent delivery pipe for high-pressure steam to the rear cylinders.

No. 600, the single 2-8-8-2 class AB, was bought as a trial in 1910. The type was enlarged and repeated a couple of years later with the AD class. It had a tractive effort rating of 94,400 pounds, about 25 percent greater than the 2-6-6-0 AA class.

Class AC was the third order of Mallets for the Virginian, numbering eight. They were last 2-6-6-0s the road bought. Their tractive effort was 90,000 pounds, not quite up to the AB's. but well above the first 2-6-6-0s. The builder photo above shows No. 517 new at Baldwin Locomotive Works in October 1910.

AC No. 514 is at Elmore, West Virginia, in the days before the electrification of this area, probably in the early 1920s. At first glance, the smallness of the locomotive makes it look almost like a rigid wheel-base machine, only the large low-pressure front cylinders calling its Mallet configuration to mind.

(all) Lloyd D. Lewis Coll.

33

No. 604, shown here new at ALCO's Richmond Works in 1912, was of class AD, the first production run of 2-8-8-2s for the Virginian, after the road experimented with the single AB class in 1910. The AC was a considerably more powerful locomotive, exerting 115,000 pounds of tractive effort in compound operation. One reason the Virginian was in the forefront of roads ordering the 2-8-8-2 was its liberal line clearances and heavy permissible axle loading. This allowed the builders to maximize the design in almost all its characteristics without the restraints that were inherent in the

Lloyd D. Lewis Coll.

clearance and weight restrictions on many older railroads. The class was the largest and most powerful locomotives of their time with a fat 100-inch diameter boiler to supply the necessary steam. These locomotives spent much of their time as pushers on the long heavy Clarks Gap grade out of Elmore, West Virginia. They outrated even the Santa Fe's large 2-10-10-2s in tractive effort.

Rev. William Stearns, H. H. Harwood, Jr. Coll.

At Elmore, West Virginia, about 1926, No. 605 rests among other large locomotives in familiar territory. The small 56-inch drivers seem disproportionately small compared with the huge boiler and firebox. Note that the tiny trailing truck seems to support little weight of the firebox, which is contained largely over the drivers of the rear engine. Sand boxes ahead of and behind the steam dome take care of helping traction for the locomotive either going forward or backward.

This Baldwin builder's photo shows the famous Triplex 2-8-8-8-4. The Virginian was attempting to maximize tractive power with this concept of putting the tender's weight to use by adding yet a third engine under it. When No. 700 was placed in pusher service up Clarks Gap and Oney Gap (east out of Princeton) grades it proved a failure because its boiler could not supply enough steam for the appetite of its six cylinders. The center cylinders received high-pressure steam, the left one exhausted into a receiving valve that supplied the front low-pressure cylinders and the right high-pressure cylinder supplied steam for use by the rear engine. Even on the short 14-mile Clarks Gap run the locomotive had to be stopped periodically to build up pressure. The Triplex was used to start heavy trains out of Princeton Yard, but even in this service it was unsatisfactory and was returned to Baldwin in 1920, where it was rebuilt into two locomotives in 1921: MD No. 410 and AF No. 610.

(Left) No. 610 represents class AF, shown in use as a hump engine at Gulf Junction, West Virginia, in 1948. The single AF was created in 1921 when the Triplex was broken apart. It was originally a 2-8-8-0 until a trailing truck was added in November 1942. It was scrapped in August 1953, and was perhaps the only locomotive in America that had three different wheel arrangements throughout its life!

Gene Huddleston

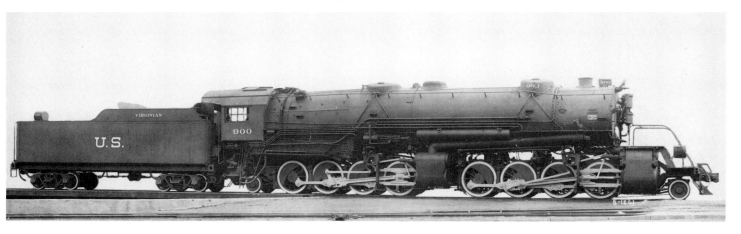

Although plainly lettered Virginian, this 2-8-8-2 built for the United States Railroad Administration (USRA) to its standard specifications was refused by the Virginian. It went to N&W as Y-3 No. 2000 in 1919. Yet the Virginian acquired its own USRA types that same year from ALCO's Richmond plant, a Southern location, which may have accounted for the rejection of the Schenectady-built No. 900.

Storming along under the wires near Princeton, West Virginia, in August 1947, is USA-class 2-8-8-2 No. 720 with a coal train. Built by ALCO's Richmond Works in 1919 to United States Railroad Administration (USRA) specifications, the 20 compound 2-8-8-2s in this class were a mainstay of Virginian coal field operations until the end of steam in the mid-1950s. During World War I the U. S. government assumed control of American railroads. In addition to running them in an effort to coordinate operations and gain efficiencies, as well as eliminate traffic bottlenecks, the USRA established a design committee to develop the best standard design that could be used on all railroads. The heavy 2-8-8-2 was one of 16 such designs, but WWI was over before any were built. However, the USRA continued to control railroads until March 1920, and many "standard locomotives" were placed in service. The Virginian liked the 2-8-8-2s and eventually ordered their own versions as well, after the USRA was only a memory.

H. Reid

Lloyd D. Lewis Coll.

No. 705 is seen here at Page, West Virginia on April 25, 1953. Originally a USA class, it was rebuilt at Princeton as a USD in June 1947, by addition of a Worthington feedwater heater and installation of a cast steel frame with integral cylinders for the low-pressure engine. It was scrapped two years after this photo was taken.

USA-class No. 706 running light at Sewells Point Yard, Norfolk, Virginia, on December 1, 1948. Pumps on the smokebox were a trademark of big locomotives on the Virginian as well as other railroads. They helped with clearances in mountainous territory. Both the C&O and D&RGW were famous for this design.

H. Reid

No. 711 comes through a natural rock tunnel at Harmco, just west of Mullens, West Virginia, on August 9, 1948. Another US class Mallet is behind No. 711 and a third is pushing at the rear, lugging thousands more tons of West Virginia's best known product to market.

A. A. Thieme

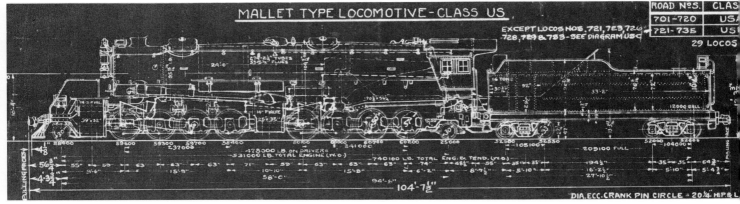

(Above) Diagram shows USA an USB class 2-8-8-2. Characteris tic of the Virginian's Mallets the tiny trailing truck wheel.

(Left) USA No. 701 with a long train of the famous Virginian "Battleship Gondolas" (116-ton coal gons—rebuilt in 1937-38 from 120-ton originals) at Carolina Junction, Virginia, on July 11, 1948.

(Below) USA No. 729 with another solid train of gons is at Norfolk in the fall of 1942.

C. A. Brown, Lloyd D. Lewis Coll.

H. Reid

(Left) USB class No. 725 working hard with a train of westbound empties somewhere west of Mullens. A scene just like this, but with doubleheaded 2-8-8-2s on the front end, is one of the author's earliest railroad memories.

William Eley

(Below) A USE 2-8-8-2 with a short train at Jazbo (New Richmond on highway maps), West Virginia, on August 18, 1952. Looks like No. 737 has just met an opposing train and is pulling eastbound out of the siding at this second station west of Elmore on the Guyandotte River Branch, which runs 45 miles down to West Gilbert and connection with both the C&O and N&W. This probably is third-class daily freight No. 98.

H. Reid

USE class No. 739 shining after coming out of Princeton shops on June 13, 1950. This class started life as USRA types built for N&W as its Y-4 class. The N&W sold them to AT&SF during World War II, and AT&SF subsequently (in December 1947) sold them to the Virginian. No. 739 had a few years left in it when this view was made, being scrapped in June 1955.

Builder's photo (ALCO-Schenectady, 1918) of Virginian Class AE 2-10-10-2. In terms of general dimensions and tractive effort, these giants would never be exceeded by any other steam locomotive except the Virginian experimental XA Triplex. They were the maximum that could be attained before the development of the superpower concept and the arrival in the mid 1920s of the "modern" steam locomotive. they exerted 147,200 pounds of tractive effort in compound mode, and could muster 176,600 pounds in simple for starting. Their front cylinders (48-inches in diameter) were so large they had to be disassembled for transportation to the Virginian, since most railroad clearance restrictions would not allow this size. They went out of general service in 1948 after a thirty-year career as the ultimate "drag-era" locomotive.

The monstrous size of the low-pressure AE cylinders is evident in this view. Because of these huge cylinders the locomotives were seldom used in simple since the boiler could not supply nearly enough high-pressure steam. Eventually this device was cut out so they could only be operated in compound. The prototype No. 800 rests under steam at Victoria, Virginia, after muscling a 16,000-ton train in from Roanoke about 1935.

H. W. Pontin, Lloyd D. Lewis Coll.

2-10-10-2 No. 807 leaves Roanoke in 1935 with a heavy coal train. This impressive workhorse generated more tractive effort than any other steam locomotive ever built (except the experimental XA class 2-8-8-8-4). But it was built to operate at very low speed with high capacities, and as such is not considered a modern locomotive. It came in the era when railroads were less concerned with speed and more with hauling capacity.

H. W. Pontin, H. H. Harwood, Jr. Coll.

This well exposed photo of AE No. 807 at Roanoke, October 8, 1937, shows to good effect the sheer size of everything about this locomotive (except its drivers, which seem puny under this massive boiler and behind those huge cylinders). Again, the most remarkable thing about the appearance of the AE is its giant low-pressure cylinders that ride entirely in front of the boiler.

Stepehn P. Davidson, Lloyd D. Lewis Coll.

Lloyd D. Lewis Coll.

No. 800 at Princeton in June 1940. This engine has just received its last major overhaul before it blew up at Stewartsville, Virginia on April, 1, 1941. It was rebuilt and served nearly two more decades before being cut up in March 1952. Following service on Clarks Gap Grade before the electrification of that line, the AEs ran mainly east of Roanoke before being succeded by the AG class Blue Ridge type 2-6-6-6s in 1945.

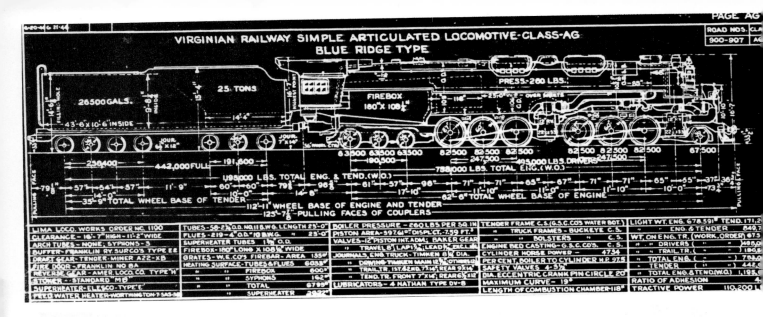

VIRGINIAN RAILWAY SIMPLE ARTICULATED LOCOMOTIVE-CLASS-AG
BLUE RIDGE TYPE

The fabulous 2-6-6-6 Simple Articulated - Class AG

Many scholars and historians who study development of the steam locomotive will acclaim the C&O H-8 class 2-6-6-6, built in 60 copies by Lima Locomotiove Works from early 1942 through 1948 as the final, best example of the Superpower concept pioneered by Lima in the 1920s. The Virginian's eight AGs (Nos. 900-907) are virtual copies of the C&O design in every aspect. Although no Virginian dynomometer reports exist, the C&O tested its locomotives repeatedly and found that they could develop 7,600 horsepower and could produce 6,600 horsepower continuously at the drawbar at 35 mph. This makes it the most powerful locomotive of all time in its ability to deliver horsepower to a train.

The Virginian came by the 2-6-6-6 in the same way as the 2-8-4. When George Brooke left the C&O's presidency in 1942 and became Virginian Chairman, he brough along Frank Beale, a C&O vice-president, to be the Virginian's president. Both men were familiar with the design work that went into the 2-8-4s and 2-6-6-6 and they believed these Superpower locomotives could do well on the Virginian's roster, lacking entirely in modern-design engines.

The AG had a rated tractive effort of 110,200 pounds, well below most of the Virginian's compound articulateds, but it was that huge boiler and monstrous firebox that allowed it to produce so much horsepower with its 67-inch drivers. Obviously not designed for slow drag work, the Virginian used the AGs east of Roanoke on coal trains, never fully realizing the tremendous horsepower potential available in these superb machines.

AG No. 907 from the rear at Norfolk, November 6, 1954, on the very eve of retiremen[t] in favor of F-M diesels. Thi[s] view shows to good advan[t]age the huge tender, larges[t] on the Virginian, as well a[s] the cavernous firebox.

Bob's Photos

AG Dimensions:
125 feet, 7-7/8-inches long over couplers; firebox 180x108-1/4-inches; tender 26,500 gallons/25-tons; Total weight, engine and tender in working order: 1,195,000 pounds; height 16 feet, 7 inches above rail head.

Two Lima builder photos of No. 900 new in 1945. The one above shows the massive, compact, powerful aspect of the locomotive, jammed with the latest appliances, and standing taller than most locomotives by at least a foot. Its high-pressure cylinders don't have the overgrown, outsized appearance of the Mallets, and its drivers seem to be in proportion to the rest of the design. The air pumps hung high and the massive pilot all give it a fearsome aspect. From the head-on view at left, one unfamiliar with modern steam locomotive design might at first be unsure what type of machine he was looking at, since the face is not that ordinarily recognized as steam's classic image. Indeed, other Superpower locomotive designs strove for a clean face, but with the AG and its sister H-8 on the C&O, the front of the locomotive was useful space to be taken up by needed appliances that couldn't be placed elsewhere on or beneath the boiler.

AG No. 900 pounds along on the flat land near Norfolk in 1947, towing the trademark Virginian gondolas.

C. A. Brown, Lloyd D. Lewis Coll.

AB No. 903 with coal at Roanoke on May 31, 1954, toward the end of its active life. The AB's were used primarily for coal trains between Roanoke and Norfolk, and the diesels are about to arrive and displace them as this photo is taken.

H. H. Harwood, Jr.

(Below) If not on its very first run, certainly an early one, No. 903 is photographed near Altavista, Virginia, with a heavy eastbound coal train in July 1945. Like the C&O, the Virginian used these great horsepower generators like drag-era Mallets, never fully exercising their potential for delivery of horsepower at speed.

George K. Shands, Lloyd D. Lewis Coll.

Section 2 - Electric Locomotives —
Classes EL-3A, EL-1A, EL-2B, EL-C

Ever since trains started climbing Clarks Gap Grade up Great Flat Top Mountain by 1909, Virginian officers faced two definitive problems: (1) the constant need for larger locomotives and cars, and better track to keep up with the transportation demand for all the coal from new mines in virtually every direction from Elmore Yard and (2) the constant complaints of its enginemen and trainmen about the near-asphyxiation they suffered in the five tunnels on that grade.

Clarks Gap Grade (which runs up the same mountain as the Norfolk & Western's Elkhorn Grade a few miles to the southwest in an adjacent county) was, of course, built single track but many of its bridges were built with towers and piers wide enough for two tracks. With several spectacular original wooden trestles replaced right away by steel bridges, the 19.2 miles of line from Elmore to MX Tower was double-tracked after World War I.

This author has discussed with several retired employees their terrible experiences of riding crouched low on the fronts of locomotives, jumping into water tanks, putting carman's cloth waste over their mouths, and finally using oxygen masks hooked to air lines—all in an attempt to merely stay alive to perform more in the company's service on "tomorrow's" Hill Runs.

In the spring of 1923, the company's board of directors authorized electrification of the 133.6 miles from Mullens up Clarks Gap and all the way to Roanoke--more than one-fourth of the entire 435-mile mainline. The huge project was phased in during 1925-26 and for the next 37 years its operation from its own powerhouse at Narrows, Virginia, improved Virginian's bottom line by millions and millions of dollars.

In fact, I believe that Virginian's freight railroad electrification was more efficient and produced more of what it was supposed to—cheap transportation translated to profits—than any similar installation in the history of Western Hemisphere railroading. That includes neighboring N&W, the Milwaukee Road in the West and anything south of the U.S.-Mexican border, but not the Pennsylvania's New York-Washington and Philadelphia-Harrisburg electrification because it was basically a passenger-hauling project.

Virginian's first of three electric locomotive classes were the 36 units generally operated in three-unit sets and nicknamed "squareheads" by those who ran them. From start of their manufacture in April 1925, a very few of these EL-3As actually ran until the end of October 1959. The last one in operation was the single unit No. 113, a work train locomotive that was in its last days parked overnight directly in front of Princeton station. Its most distinguishing feature was its individualized nickname, the word "Clarabelle" spelled out by whimiscal shopmen in white paint on one of its small pilots.

The image of these world's-largest locomotives humming up and down mountains with a mile of loaded hopper cars came to symbolize what the general public thought of as "modern railroading" in the mid-1920s when the inauguration of Virginian electrified train service brought the little Class I some national media attention.

Coal shipments increased over the years and the EL-3As ran for more than 20 years before management considered adding to the fleet. This they did with only four double-unit EL-2Bs manufactured by General Electric at Erie, Pennsylvania, numbered 125-128, and dubbed "streamliners" by their crews. And they truly were streamliners, a one-class departure from the otherwise utilitarian design of all other Virginian locomotives of all shapes and sizes.

A sleek style carbody reminiscent of EMD's F-units was built over the tough insides which weighed more than one million pounds per two-unit set and produced 6,800 horsepower.

Practicing scholar and "boomer" railroad president John W. Barriger, one of the most quoted and interesting officials in mid-20th Century American railroading, had praise for the technology of Virginian's final class of electrics, the 12 GE 3,300-horsepower EL-C class "rectifiers" built in 1956-57. Virginian historian H. Reid had a simpler description for them: he said they had the beauty of a brick.

Regardless, the 130-141 performed well, enabling the retirement of the 30-year old squareheads.

Subsequent New Haven, Penn Central, and Conrail service then added to their fame until they were all cut up except Nos. 130 (now in a museum in Connecticut), 133 (now in an Elkhart, Ind. museum), and 135. No. 135 is in Virginia Museum of Transportation at Roanoke, in company with SA 0-8-0 No. 4—one from the first order actually placed under the Virginian name, and one from the last, by the sheerest of coincidences.

Electric Locomotive Specifications								
Class	Road No.	Builder/Date	Motors	Wt. On Dr.	Total Wt.	T. E.	Horsepower	Max. Speed
EL-3A	100-111	ALCO/Westinghouse	6	922,580 lbs.	1,282,380 lbs.	231,000	7,125	38 mph
EL-1A	110-115	ALCO/Westinghouse	6	309,300	430,560	77,000	2,375	38 mph
EL-2B	125-128	General Electric	16	1,033,832	1,033,832	260,000	6,800	50 mph
EL-C	130-141	General Electric	6	394,000	394,000	98,500	3,300	65 mph

Notes:
EL-3A wheel arrangement 1-D-1 in each of three permanently coupled units
EL-1A wheel arrangement 1-D-1, single units, not permanently coupled
EL-2B wheel arrangement B-B+B-B in two semi-permanetly coupled units
EL-C wheel arrangement C-C in individual units

Lloyd D. Lewis Co

ALCO builder's photo of the first Virginian electrics, class EL-3A, new in May 1925. The three-unit locomotive was permanently coupled and this is No. 100, the first of 12 similar sets that established the Virginian's electric roster. In the 1920s many steam railroads were considering the efficiency and cleanliness of electricification of their heaviest freight lines, but only a few ever attempted any large scale conversion, notably the Virginian, Norfolk & Western and Milwaukee. The Pennsylvania's electrification was largely in passenger operations.

Starting tractive effort of the EL-3A was 231,975 pounds at 25% adhesion and 277,500 pounds at 29.9% adhesion. In hourly operation the tractive force of these brutes was 162,000 pounds at 14.1 mph and 94,500 at 28.3 mph. The continuous rating was 135,000 at 14.2 mph and 78,750 at 28.4 mph. As can be seen these locomotives could far out-perform the biggest steam on the line, and maintenance was much easier. There was no need for large fueling and water stations, or roundhouses for constant repair, laborers for fire tending/cleaning, etc. Of course the Virginian didn't realize all these savings because it operated steam in the same territory for branches, passenger operations, and some mainline freight.

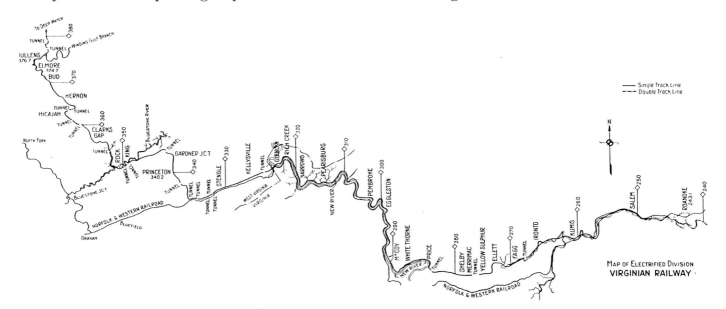

Map of the Virginian electric territory as first installed in 1924-26. See Page 50 for the profile of this line. The Virginian was never a line to shrink from the superlative. Built to the best and most modern engineering standards, with ample clearances and axle loading ratings, it early on established the Mikado type, helped establish the Mallet in road service, built the biggest, heaviest, and most powerful locomotives of its time, and experimented with a Mallet of Triplex design. So it was no wonder that it would and could, given its vast resources from the booming coal business, undertake one of the world's largest electrification projects. It paid Westinghouse $15,000,000 to install the wire and all electrical apparatus including a 60,000 Kw steam turbine generator plant at Narrows, Virginia.

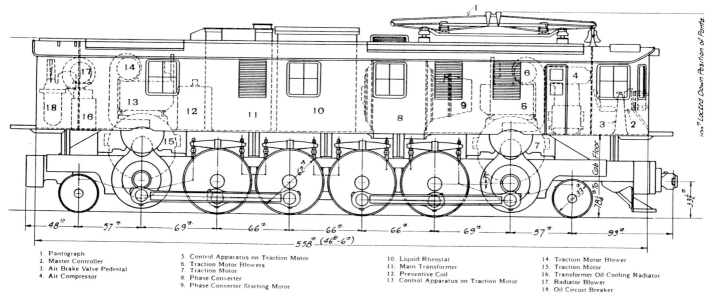

1. Pantograph	5. Control Apparatus on Traction Motor	10. Liquid Rheostat	14. Traction Motor Blower
2. Master Controller	6. Traction Motor Blowers	11. Main Transformer	15. Traction Motor
3. Air Brake Valve Pedestal	7. Traction Motor	12. Preventive Coil	16. Transformer Oil Cooling Radiator
4. Air Compressor	8. Phase Converter	13. Control Apparatus on Traction Motor	17. Radiator Blower
	9. Phase Converter Starting Motor		18. Oil Circuit Breaker

A Westinghouse drawing from 1925 shows the components of the new EL-3A locomotive unit. From the photos and this drawing the reader can easily see that the traction motors were mounted in the body of the unit, transmitting their power to a "jack shaft" that then turned the driving wheels by means of a connecting rod in much the same way as a steam locomotive. Later electrics had much smaller drivers and placed the traction motors directly around their axles.

EL-3A No. 103 is getting its sandbox filled at Mullens Motor Barn in 1925, just after completion of the electrification project. The boxy electrics were certainly something new in coal country when compared with the smoky, noisy steam locomotives they supplanted. So new, in fact, that the locals called them "motors." The Mullens repair shop was called "Mullens Motor Barn" and the single-track Roanoke version called the "Roanoke Motor Shed," much more interurban/trolley terms than would be expected in coal country.

47

An EL-3A, No. 103, takes a train across the trestle over W. Va. Route 10 at Garwood, West Virginia, about 1950. This "Hill Run" is curving toward the first tunnel on Clarks Gap Grade.

This low-angle view gives some appreciation for the size and power of EL-3A No. 101, its pantographs standing high to reach the wires adjacent to "Beanery Hollow" at Elmore, West Virginia, on March 14, 1956.

(Above) An eastbound coal train passes the small depot at Narrows, Virginia, with EL-3A No. 101 for power in 1948. The N&W and Virginian both pass through a steep Appalachian gap at Narrows, one on each side of the New River. *(Below)* No. 106 heads an eastbound extra coal train at Kellysville, West Virginia, on June 10, 1949. The typical railroad wooden water tank scene is cluttered by the electrical substation.

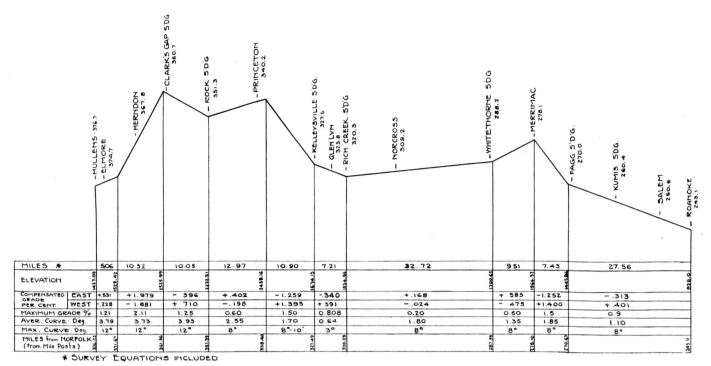

MILES *	5.06	10.32	10.05	12.97	10.90	7.21	32.72	9.51	7.43	27.56
ELEVATION	1417.03	1529.42	2515.99	2232.51	2498.16	1674.13	1524.96	1790.65	1964.37	1445.84	926.0
COMPENSATED GRADE PER CENT — EAST	+.531	+1.979	−.596	+.402	−1.259	−.340	+.168	+.585	−1.252	−.313
COMPENSATED GRADE PER CENT — WEST	−.228	−1.681	+.710	−.198	+1.395	+.391	−.024	−.475	+1.400	+.401
MAXIMUM GRADE %	1.21	2.11	1.25	0.60	1.50	0.808	0.20	0.60	1.5	0.9
AVER. CURVE Deg.	3.79	3.73	3.93	2.55	1.70	0.64	1.80	1.35	1.85	1.10
MAX. CURVE Deg.	12°	12°	12°	8°	8°-10'	3°	8°	8°	8°	8°
MILES from NORFOLK (from Mile Posts)	376.67	371.61	361.36	351.33	338.40	327.49	320.29	287.53	278.10	270.67	243.11

Station profile labels: MULLENS 376.7 · ELMORE 374.7 · HERNDON 367.8 · CLARKS GAP SDG 360.7 · ROCK SDG 351.3 · PRINCETON 340.2 · KELLEYSVILLE SDG 327.5 · GLEN LYN 323.8 · RICH CREEK SDG 320.3 · NORCROSS 309.2 · WHITETHORNE SDG 286.2 · MERRIMAC 278.1 · FAGG SDG 270.0 · KUMIS SDG 260.4 · SALEM 250.6 · ROANOKE 243.1

* SURVEY EQUATIONS INCLUDED

Condensed Profile of the Electrified Section

(Above) Profile of the Virginian's electrified territory between Mullens, West Virginia, and Roanoke, Virginia. The Clarks Gap grade out of Elmore Yard is the steepest on the main line and is against eastbound loaded coal traffic.

(Right) No. 103 at Alpoca, West Virginia, in 1950. The Hill Run flies the white flags of an extra train on this double track, train-order, telegraph and telephone-dispatched mountain railroad. The power director at the Narrows Power Plant controlled distribution of the 88,000 volts of electricity that was stepped down to 11,000 volts where the locomotive pantograph rubbed the trolley contact wire. A coal company store is just beyond the grade crossing to the left of the track. Around the mountain to the right (about where the EL-3A pusher is shoving) is the Alpoca coal mine, which is just about the only reason to name a place in West Virginia.

(Right) No. 100, the prototype EL-3A (but running only two "permanently coupled units today), at Oakvale, West Virginia, in October 1948, with a train of empty uniform hopper cars. As can be seen in the pages of this book, the Virginian was a railroad of many bridges on its western end, traversing the mountains and ravines of West Virginia coal country.

(Below) Emerging from the tunnel through Clarks Gap at Algonquin, West Virginia, in September 1948, EL-3A No. 104 carries the white flags of an extra. These machines made a steam locomotive-like clank-ing noise as their side rods cranked their drivers along, not quite in uni-son. The top-mounted and polished bell and trimmed windows gave a little decoration to the front of these otherwise quite utilitarian locomo-tives. The catenary pole at right is the 952nd pole west of Narrows Power Plant and is on the north side of the track.

(both) B. F. Cutler, H. H. Harwood, Jr. Coll.

51

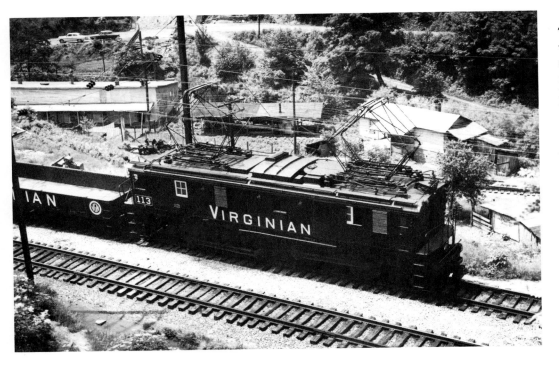

The EL-1A locomotives were built to operate as a single unit and arrived new in February and March 1926. In September of that year Nos. 110, 111, and 112 were rebuilt into a single unit EL-3A with Number 110, while Nos. 113, 114 and 115 became No. 111. No. 111 was again converted to single units with old Nos. 113, 114, and 115 in January 1927, only to be rejoined once more as No. 111 in December 1941. This combination was apparently again broken up as single units before it was scrapped in 1960. Here No. 113 is on a work train at Matoaka, West Virginia, on June 13, 1956. As a single locomotive the El-1A could develop 2,375 horsepower and tractive effort of 77,000 pounds, so they were a formidable locomotive even in single. This unusual overhead view affords a good appreciation of the pantographs and roof.

An eastbound coal train with No. 103 for power, crosses over U. S. highway 219-460 at Oakvale, West Virginia, on September 4, 1953. The sign beside the road advertises the Virginian Hotel in Princeton.

Squarehead EL-3A No. 101 meets streamliner EL-2B No 126 (see following pages) at Kelleysville, West Virginia, at 4:25 p.m. June 10, 1949. A perfect juxtaposition of the old and new, the undesigned and the highly styled. Note the crewmen lounging in the shadow under the water tank on the right.

EL-3A No. 101 at Roanoke in September 1949. Below the body these locomotives looked a lot like a steamer, with counterbalanced driving wheels connected with mainrods, and when they ran they clanked like a steam locomotive as well.

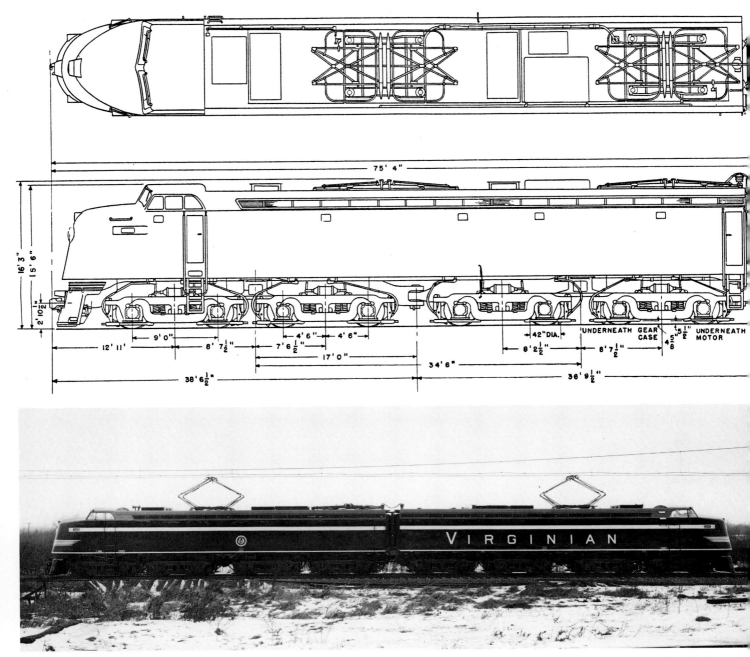

Dimensions shown on diagram: 75' 4", 16' 3", 15' 6", 2' 10½", 9' 0", 12' 11", 8' 7½", 7' 6½", 4' 6", 4' 6", 17' 0", 34' 6", 38' 6½", 42" DIA., 8' 2½", 8' 7½", 36' 9½", UNDERNEATH GEAR CASE, 5⅛", UNDERNEATH MOTOR, 4⅝"

VIRGINIAN

Lloyd D. Lewis Co

The EL-2B paired electrics were known, for obvious reasons, as the "streamliners." The locomotive consisted of these two permanently coupled units, which weighed 1,033,832 pounds and all of it resting on the drivers with traction motors slung across the axles of the eight powered trucks. These sleek locomotives with their VGN emblem in winged stripe on the nose were certainly the very antithesis of the image of the EL-3As which had absolutely no styling. The EL-2Bs arrived in January 1948, had a 6,800 horsepower rating, and were used on a variety of heavy work until the end of the Virginian. After the N&W merger in 1959 they were renumbered as individual units (N&W 221-228) and remained until the wires came down in 1962. The diagrams above and on the facing page are from General Electric, the builders of these fine locomotives, and illustrates well their clean lines, as does the broadside portrait below.

According to General Electric literature of the day, these locomotives were designed specifically for heavy mountain work. This same pamphlet stated that if the horsepower rating were calculated in the same way as diesel horsepower, the EL-2Bs would be rated at 8,000. The rigid wheel base of each truck was only nine feet, and all trucks were exact duplicates, with all connections designed to be broken with minimum effort so that they could be interchanged in a very short time. The locomotive was powered by two GE-746 d-c series-wound traction motors, one on each axle.

54

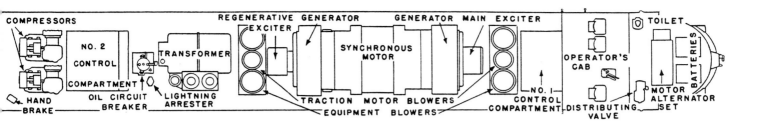

COMPRESSORS REGENERATIVE GENERATOR GENERATOR MAIN EXCITER TOILET

NO. 2 CONTROL COMPARTMENT EXCITER SYNCHRONOUS MOTOR OPERATOR'S CAB BATTERIES

TRANSFORMER

HAND BRAKE OIL CIRCUIT BREAKER LIGHTNING ARRESTER TRACTION MOTOR BLOWERS NO. I CONTROL COMPARTMENT DISTRIBUTING VALVE MOTOR ALTERNATOR SET

EQUIPMENT BLOWERS

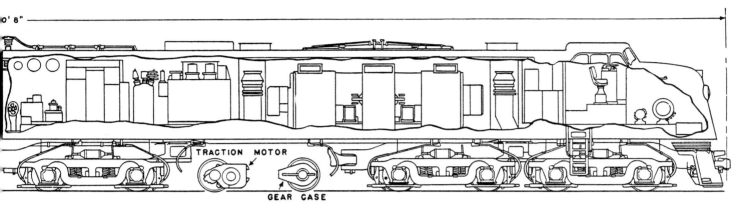

0' 8"

TRACTION MOTOR

GEAR CASE

(Above) This GE diagram shows the layout of the EL-2B and a cutaway elevation, identifying each major component.

(Left) Head-on view of an EL-2B gives a sleek, streamlined appearance, with the VGN logo centered in swept striping. A handsome face, indeed.

(Below) Another delivery portrait showing the locomotive at an angle. The GE industrial styling designers did an excellent job with this locomotive.

(all)GE, Lloyd D. Lewis Coll.

Still lettered Virginian, but with N&W numbers 227-223 after the merger, with eastbound coal train at Princeton Yard; shops in the background. Princeton was the Virginian's principal shop for steam and diesel work, but electric locomotive work was primarily the responsibility of the shop at Mullens, West Virginia.

No. 128 at Princeton May 19, 1954. The giant streamliner appears to be headed westbound with Time Freight No. 71 rolling through the yard at the west end of Princeton Shop between the system storehouse and the large black metal carshop. No. 71 is either setting off or picking up cars here or is meeting an eastbound on the main line, which swings around the base of the hill west of the shops and yard.

William Swartz, Lloyd D. Lewis coll.

No. 126 has an eastbound manifest freight train crossing the East River bridge just west of Glen Lyn, Virginia, in August 1948. The Virginia-West Virginia state line marker is near the right edge of the photo and No. 126 is about to cross over the N&W main line.

B. F. Cutler, H. H. Harwood, Jr. Coll.

Rumbling over the steelwor[k] of Virginian's longest bridg[e] (2,155 feet) and breaking th[e] lull of the late spring evenin[g] EL-2B No. 126 is eastboun[d] with coal at 6:15 p.m. o[n] June 12, 1956. These co[n]crete piers (and one more [to] the left of the photo) were th[e] tallest in the world when com[m]pleted in 1909. Today on[ly] the piers stand, Virginian['s] line at this point remove[d] about 20 years ago becau[se] of a highway constructi[on] project. The N&W main lin[e] visible above the river at le[ft] passes under the bridge rig[ht] behind the trees.

Richard J. Cook, Sr.

Single unit No. 227 (N&W number) parked, with pantographs down at N&W's Shaffer's Crossing engine terminal in Roanoke in the summer of 1963, awaiting scrap.

Lloyd D. Lewis Coll.

No. 127 arriving Roanoke Yard August 4, 1956, with coal train eastbound. The Virginian was not bashful about its name as evidenced by the huge letters on all its coal cars.

Steve Patterson

58

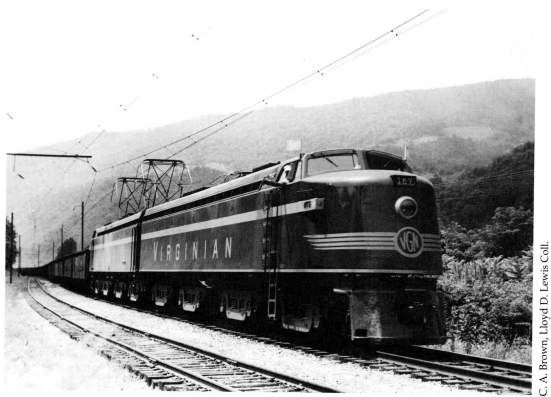

Westbound No. 127 with a solid train of the Virginian's famous "battleship gondolas" at Rich Creek, Virginia, on July 13, 1948, in its first year of operation. Only the rear pantograph of each unit is up. Note the metallic extra flags.

C. A. Brown, Lloyd D. Lewis Coll.

Richard J. Cook, Sr.

At Kellysville, West Virginia, No. 126 carries coal empties westward at 4:15 p.m., June 10, 1949. See photo of this train meeting Extra 101 East here on page 53.

The Virginian bought class EL-C rectifier electrics from GE in late 1956 and early 1957, numbering them 130-141. They were geared for 65mph, the fastest of any Virginian electric, and could develop 98,500 pounds of tractive effort. Better looking than the box cabs of 1925, they were no match for looks with the streamlined EL-2Bs. In this General Electric builder photo the unit hasn't received its road number yet, but is probably No. 130. This photo, like many other electric and diesel locomotive builder's pictures, was taken on GE's test track, officially the East Erie Commercial Railroad at Erie, Pennsylvania.

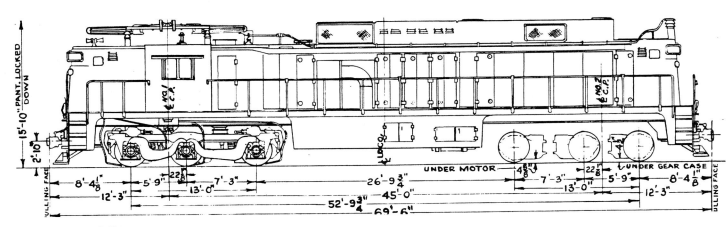

Lloyd D. Lewis Coll.

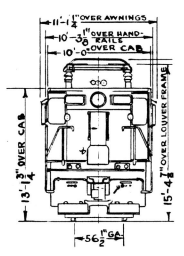

Diagrams of the EL-Cs show principal dimensions and configuration. The N&W renumbered the units to 230-241 in 1960, and then sold them in 1963 to the New Haven Railroad. The Virginian EL-Cs allowed the railroad to retire the aging, clanking, boxy squarehead EL-3As. But they, too, outlived their usefulness, if not their mechanical and electrical capabilities, when last-owner Conrail shut down all its electric freight service in 1976. The ex-Virginian 131 went to a Connecticut museum, No. 133 went to an Elkhart, Indiana museum, and No. 135 came back to Roanoke for the Virginia Museum of Transportation. Painted in a full original Virginian paint scheme, No. 135 was presented as a museum piece in front of the Roanoke passenger station with long-time engineer Kent Womack posing for photos in the cab.

New and as yet unnumbered at the GE plant in Erie, Pennsylvania, EL-C No. 130 strikes a dramatic pose. It is on the five-mile test track at the plant for final testing, probably in late October 1956. It's interesting to note the multiplicity of rails in the track, since GE built locomotives to many different gauges!

No. 230 (N&W number) and two mates at the Mullens Motor Barn after Virginian had passed into oblivion. This was during the 2-1/2 years that N&W ran these units in sets of three instead of the Virginian standard of two. Train lengths increased, naturally.

(both) Lloyd D. Lewis Coll.

N&W number 233 (formerly VGN 133) rests at Mullens, West Virginia, at the west end of electric territory on October 14, 1960.

Nos. 132 and 133 work an eastbound coal train just west of Salem, Virginia, in March 1958, near the end of their run at Roanoke, the eastward limit of electrification. Diesels will muscle the train on to Sewells Point piers in Norfolk.

An eastbound coal train has two EL-C electrics for power as it rolls through a cut east of Princeton, West Virginia, en route to the sea, in June 1958. This cut leads to Oney Gap Tunnel, the first grade east of Princeton and once a test area for the XA experimental Triplex. Oney Gap is named for the Oney family of Princeton, who still live there and under whose farm the tunnel is bored. Herb Harwood shot this photo from a wooden overhead bridge. The dirt from this cut was used in building a fill for Princeton Yard.

(both) H. H. Harwood, Jr.

No. 139 leads a mate with westbound merchandise train No. 71 leaving Roanoke in June 1958. The scene here is near the west end of Roanoke Yard on the bank of the Roanoke River. Just under the bridge from which the photo was taken is AG Tower, a major track scale weighing location for eastbound coal trains.

Lloyd D. Lewis Coll.

This N&W official photo depicts one of the Virginian EL-C types in full N&W black paint and lettering at Roanoke. No. 235 was Virginian 135, and was the only Virginian electric locomotive ever painted in the black scheme. No. 135 was involved in a sideswipe derailment at Kumis, Virginia, in November 1958 and was still out of service when the N&W merger occurred.

The EL-C class was much traveled after the end of the Virginian, going, of course, first to the N&W, then to the New Haven, and finally to Penn Central and Conrail. Shown here at Orangeville engine house, Baltimore, Maryland, in November 1972, two units await another freight run. These units sat in the South Roanoke Yard for more than a year before they were sold to the New Haven for $20,000 each, much less than their purchase price of just six years before. Nine of the original 12 EL-Cs were scrapped by General Electric after their return as trade-ins on new diesels by Conrail.

Norman Perrin

Section 3 - Diesels: Big and Standard, With One Exception

Following a survey of many railroads and builders, Virginian did what no other road its size ever did: buy virtually all its diesels from Fairbanks-Morse. F-M had the relatively large size of locomotive that Virginian was used to in one unit: the 2,400-horsepower aptly-named "Trainmaster."

The 25 TMs were the largest order of that model that Fairbanks-Morse produced and they mastered the coal and freight trains on Virginian coal branches and the mainline west of Elmore during the last five years of the railway's existence. On until the American Bicentennial on July 4, 1976, at least one of them was in service, so N&W got Virginian's money's worth out of those units.

They were based for maintenance at Mullens Motor Barn, where General Foreman Randy Hearn and his men gave lots of tender loving care to these beasts that emitted clouds of blue smoke. These units were commonly used on Hill Runs but usually stayed west of Clarks Gap. They also were found on all lines west of there and were the standard Elmore Yard engines for 20 years.

The most famous of the Trainmasters was No. 171 (ex-Virginian 71) which after N&W took over the property was "permanently" based at Oak Hill, West Virginia. From there it worked the coal mine jobs on short branches to Summerlee and Lochgelly, supplied the jointly served (with C&O) mine at Macdonald and hauled loads to and brought back empties from the N&W mainline at Oak Hill Junction, all in Fayette County, West Virginia. This unit was individualized by the electric hookup from a power company pole at the Oak Hill depot to a heater in No. 171's engine which allowed the crew to keep vital parts warm on those cold, cold West Virginia nights. Just like a diesel-powered automobile, this idea resulted in quicker engine starts in the morning—and also the nickname "Plugged-in Trainmaster."

F-M's 1,600-horsepower four-axle model (not a "Baby Trainmaster," which also had 1,600 horsepower but six axles) was chosen by Virginian management to replace the AGs and BAs on the East End and handled thousands of coal trains from Roanoke to Sewells Point until the merger in 1959 and for several years thereafter. They were most photographed as yard engines at N&W's Lamberts Point coal terminal, shoving thousands of N&W (and Virginian) loaded hoppers up onto the famous Pier 6 and others nearby for export and coastwise coal shipping.

Two locomotives—Nos. 15 and 16—were equipped with steam generators so Virginian officers could use them on their semi-annual inspection trips. These generally included a 200-series steel rider coach for the train crew and two or more office cars, like the *Guyandotte River* and *Fairhaven*, named incidentally, for Virginian builder H. H. Rogers' hometown of Fairhaven, Massachusetts.

In addition, Nos. 48 and 49 were built late in the Virginian's life—in October 1957—because Nos. 23 and 28 were demolished in a head-on collision at Huddleston, Virginia—a misspelling, furthermore, of Mr. Rogers' own middle name—earlier that year. The 40 DE-S class diesels were second only in number to the total of 42 MB 2-8-2s: once again, good designs at both beginning and end of Virginian motive power history.

Virginian was so conservative even in those prosperous mid-century years that when time came to retire SA 0-8-0s Nos. 2 and 4 (which alternated on a light-duty industrial line connecting Virginian and Seaboard Air Line at Suffolk, Virginia), management wouldn't buy another DE-S. This would be a special little locomotive for a special job. So in July 1954 Virginian purchased a 13-year-old second-hand General Electric 44-tonner with only 380 horsepower, numbered it 6 and called it class DE-SA. Its career at Suffolk also ended with the N&W merger and it was sold again in 1960.

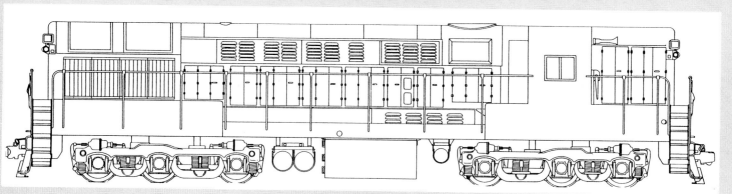

Diesel Sepcifications						
Class	Type	Road No.	Builder/date	Horsepower	Wt. on Drivers	Tractive Effort
DE-SA	B-B	6	ALCO/GE/1941	380	88,400 lbs.	22,100 lbs.
DE-S	B-B	10-49	F-M/1954-57	1,600	262,200 lbs.	65,500 lbs.
DE-RS	C-C	50-74	F-M/1954-57	2,400	394,500 lbs.	98,625 lbs.

This excellent photo shows No. 11, a member of the DE-S class, new, fresh at Fairbanks-Morse in June 1954. The F-M locomotives were unusual in that their prime movers were of "opposed piston" construction, where two pistons were juxtaposed against each other in the same cylinder and combustion drove them in opposite directions. The design was never popular and Fairbanks-Morse never became a major force in the road diesel locomotive business. It built only 445 road switchers total, of which 241 were of the H16-44 variety such as the Virginian's Nos. 10-49.

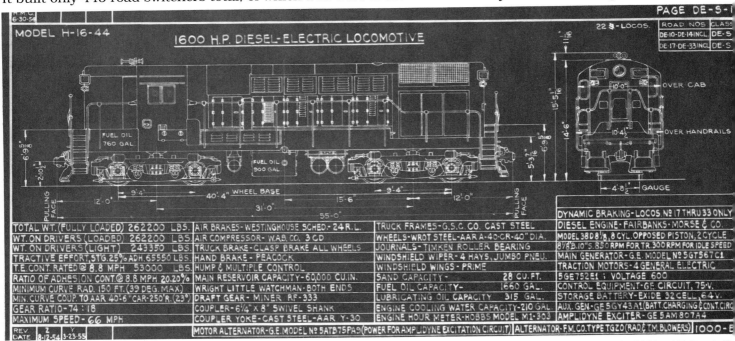

(both) Lloyd D. Lewis Co

The Virginian's official diagram showing the 1,600-hp F-M diesel locomotive of which there were 40 on the roster. Although these units were sometimes confused with the F-M's "Baby Trainmasters" the latter had six axles, whereas the Virginian units rode on four-axle trucks, though they had the same 1,600 horsepower rating. When Virginia decided to dieselize its steam operations, it decided to go with a single builder, and it got only two models, this 1,600 horsepower unit, and the larger 2,400-horsepower Trainmaster. The latter were used primarily on the West End where they were ideally suited to mine runs where heavy loads and heavy grades were the rule, whereas the DE-class was used mainly east of Roanoke to supplant the BA and AG classes that held sway here in the last decade of steam.

(Above) At Victoria in 1955, No. 24 runs light through the yard.   It seems out of place with the old wooden station and its loungers. The 1,600 hp F-M locomotives were used primarily east of Roanoke on Virginian's Norfolk Division to Sewells Point.

(Below) A pair of DE-S units power a coal train east out of Roanoke at JK Tower on August 6, 1958, paired with long hoods back to back. Though rare and not particularly good-looking, Fairbanks-Morse locomotives have become sort of mythical in the railfan and railroad history community. They looked as good as they did anywhere in the snazzy Virginian livery.

No. 26 switches a cut of cars in the middle of Roanoke Yard in April 1956.

Fairbanks-Morse's model number for the 1,600 hp locomotives was H16-44. Here Nos. 20 and 39 handle a westbound train at Victoria, Virginia, in March 1956, with capabilities that exceeded most steam locomotives previously assigned to this run.

Fairbanks-Morse took this builder's portrait view of its "Trainmaster" model, newly painted and ready for delivery to the Virginian in April 1954. The ALCO PA-model has been called an "honorary steam locomotive" by many in the railroad enthusiast community, and perhaps that would apply to the F-M Trainmasters as well. They have become a much discussed locomotive, but apparently only one has been preserved—a Canadian Pacific TM in a museum in Western Canada.

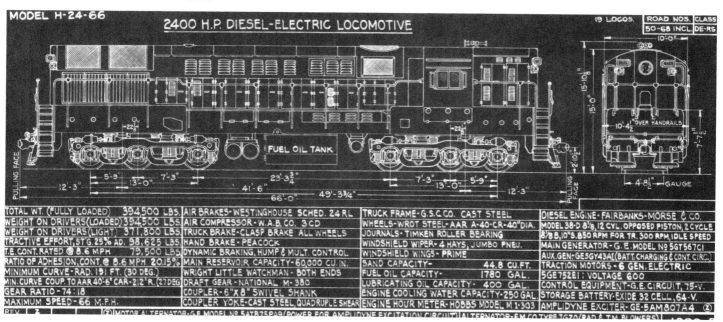

(both) Lloyd D. Lewis Coll.

Virginian diagram depicting the Trainmaster design (F-M Model H24-66), which the railroad classed DE-RS. F-M built a total of 107 of its Trainmaster model, so the Virginian's 25 accounted for almost one-fourth of the model's entire production.

Brand-new Fairbanks-Morse Trainmaster No. 59 with DE-S No. 12 on westbound empty train of 116-ton gondolas at Victoria July 16, 1954. The officials (Trainmaster Tommy J. Nichols, Jr., Road Foreman of Engines Jimmie Williams, Norfolk Division Superintendent J. P. Strickland, and Claim Agent L. G. Walker) climbed off the units to pose in front of Virginian's newest motive power.

(all) Lloyd D. Lewis Coll.

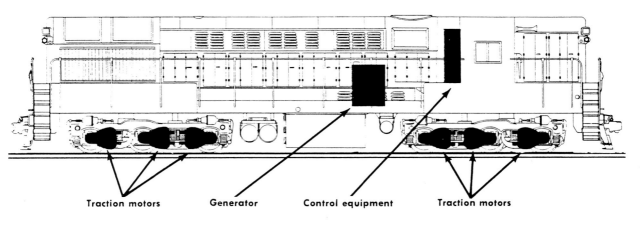

Traction motors   Generator   Control equipment   Traction motors

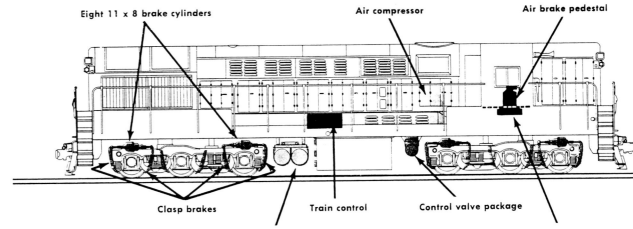

Eight 11 x 8 brake cylinders   Air compressor   Air brake pedestal

Clasp brakes   Train control   Control valve package

Two diagrams lifted from an F-M manual on the "Trainmasters" illustrate the air brake systems (top), and the traction motors, generator and controls.

While being delivered to Virginian from the Fairbanks-Morse factory in Beloit, Wisconsin, several Trainmasters were numbered temporarily in the 800 series, maybe because of possible conflicts with numbers of existing locomotives on delivering carriers. Thus in March 1954, two TMs are on their way to Elmore for the first time. Location is not known.

Lloyd D. Lewis Coll.

Peg Dobbin, H. H. Harwood, Jr. Coll.

F-M H24-66 No. 58 on a mine run shifter train at Royal, West Virginia, up the Winding Gulf Branch, in April 1955. The scene is typical of Virginian's coal territory with the large mine tipple and its many tracks for loading, hard by the small shanties occupied by miners. Up until a year before, this type of duty would have been performed by a 2-8-8-2 Mallet. The F-M road switchers were versatile and powerful enough to handle this type of business with much less fuss than the steamers did.

Curving around the Slab Fork, West Virginia trestle, two F-M Trainmasters muscle a train of heaped coal eastward out of a typical West Virginia "hollow" in April 1955, soon after replacing steam in this service. The Caperton Coal Co. town of Slab Fork is under both sides of this bridge. Off to the right is the local coal mine branch, which leads to one of the very first mines ever opened for coal shipments in Virginian country.

(both) Peg Dobbin, H. H. Harwood, Jr. Coll.

A Trainmaster takes some interchange traffic across the Kanawha River bridge at Deepwater, West Virginia, in 1955. It was at Deepwater that the Virginian began as a shortline connecting with the C&O. The bridge here connected it with the New York Central's Kanawha & Michigan on the north side of the river.

Virginian Trainmaster No. 51 switches a cut of coal loads at the east end of Elmore Yard in April 1955. In the background are the steam-era concrete coal dock and the rectangular Elmore Steam Shop, both falling to disuse with the arrival of new yellow-and-black diesels. The Guyandotte River is over the bank in the foreground.

DE-RS with a coal train near Sophia, West Virginia, in April 1955. Paralleling the C&O's Winding Gulf Subdivision (at left), a new Trainmaster hauls coal on Virginian's Winding Gulf Branch. The F-Ms generally were used one at a time on all the coal branches and were usually double-headed in multiple-unit control on mainline coal and time freight trains.

(both) Peg Dobbin, H. H. Harwood, Jr.

The last of its class, DE-S No. 149 (with N&W number) sits on a turntable lead at Sewells Point Yard in Norfolk about 1960. The Virginian diesels were arranged to run long hood forward much the same way that neighbor N&W operated, whereas most railroads ran the short hood and cab forward for better visibility. The long-hood-forward operation did offer better collision protection for the crew.

(Below) In N&W black, No. 156 works a string of N&W cars as the East End Elmore Yard switcher job in 1964. Virtually all Virginian diesels still existed at this date, but by January 1, 1975, all were off N&W's roster except two Trainmasters (Nos. 171 and 173) and five 1,600-hp DE-S units.

Neil Boggiano

Jim Shaw

(Above) DE-S No. 40 on Roanoke Yard about 1958. The Virginian paint diesel scheme was a drastic departure from what one would have expected from so conservative a steam road. The body was bright yellow with a black band around the middle and large VIRGINIAN capital letters on the side. For practically everything the road lettered it used the very large gothic letters. In addition, the black-and-yellow striped end pilots were outsstandingly visible and would have been perfect as a good example of a 1990s "Operation Lifesaver" paint scheme.

(Right) This overhead view shows TM No. 66 switching Time Freight No. 72 as it holds down the East End Elmore switcher job. The daring photographer clings to a multi-story yard light tower for this dramatic shot. East End Elmore Yard office hangs onto the bank of the Guyandotte River at top left. This structure was demolished in favor of a new yard office in 1991. The west switch of Clarks Gap Double Track is at top right.

Jim Shaw

75

Mallory Hope Ferrell

The light connecting line between Virginian and SAL at Suffolk, Virginia, allowed replacement of its 0-8-0 switchers by something less than the 1,600-hp F-M units that were Virginian's standard, so it bought a second-hand GE center cab 44-ton industrial switcher and gave it the DE-SA designation and road number 6. It worked out of Suffolk until it was sold to Eastern Gas and Fuel Associates in 1960 after the N&W merger. Here, in the rain, it has SAL box cars in tow at Suffolk.

Bob Lorenz

Posing proudly at Suffolk just after its arrival in 1954, is No. 6, with the large Planters Peanut factory in the background.